DINOSAUR DOCTOR

DINOSAUR DOCTOR

THE LIFE AND WORK OF GIDEON MANTELL

EDMUND CRITCHLEY

AMBERLEY

First published 2010

Amberley Publishing Plc
Cirencester Road, Chalford,
Stroud, Gloucestershire, GL6 8PE

www.amberley-books.com

British Library Cataloguing in Publication Data.
A catalogue record for this book is available from the British Library.

ISBN 978 1 84868 947 3

Typesetting and Origination by Fonthill.
Printed in the UK.

CONTENTS

INTRODUCTION

DR GIDEON MANTELL LLD, FRCS, FRS: FROM DINOSAUR TOOTH TO DISTORTED SPINE

Professor George J. Stigler, the Nobel Laureate economist, is also known for his law of eponymy, which says that no scientific discovery is named after its original discoverer. This law can be applied to the career of Dr Gideon Algernon Mantell, responsible for the discovery of the Iguanodon and six other dinosaurs, for his depiction of the Age of Reptiles moving from a marine to a land-based environment, and for his numerous additions to our knowledge of palaeontology and geology. In his lifetime, jealous contemporaries took advantage of the misfortunes of his family, his precarious financial position and persistent ill-health (particularly from the time of his accident in October 1841) to undermine, re-assign and minimise acknowledgement of his wide-ranging researches. Unsurprisingly, the biography of the *Dinosaur Hunters* by Deborah Cadbury (1988), a producer of TV documentaries, was subtitled 'A true story of scientific rivalry and the discovery of the prehistoric world'.

Mantell's ephemeral attributes are forgotten. As a lecturer he was outstanding. As a writer his books were popular, instructive and scientific, but unfortunately can now only be found among the dust on the least-used shelves. His exceptional museum collection of fossils and antiquities was bought for a pittance, then scattered and distributed among the other exhibits of the British Museum. Today he is remembered as much for his distorted spine – for many years a prize exhibit of the Royal College of Surgeons' Hunterian Museum – as for his pioneering scientific reputation in the early part of the nineteenth century.

Fortunately it is possible to reconstruct his life and works by taking advantage of his writings; his private journals covering thirty-four years of his life, and his letters to friends and fellow scientists at home and abroad have been utilised by his scholars. An early biographer, dental surgeon Sidney Spokes (1927), came to live in the house where Mantell had opened his medical practice and made particular use of his correspondence with Professor Silliman of Yale. E. C. Curwen (1940) abridged the four volumes of his journals to describe the scene in Mantell's own words. Dennis R. Dean (1999) has meticulously researched the previously unutilised manuscripts and documents at the Alexander Turnbull Library, Wellington, New Zealand, as well as the lives and writings of his contemporaries, concentrating on his scientific achievements and the vagrancies of his family, within Sussex, where Mantell lived for most of his life. More recent contributors such as Colin Brent, the historian of the town of Lewes, and Anthony Brook, a geologist with an interest in antiquities, have provided additional facts concerning his life and works.

His discoveries fit into a contemporary picture of a time when people were obsessed with former worlds, the biblical story of the creation of life, the deluge, and past extinctions, preceding theories of mutability and evolution. There was an expansion of literacy and a search for education and knowledge. Lending libraries thrived, people attended lectures at the Mechanics Institute, scientific societies came into being: the Linnaean for botany, the Geological, the Microscopical, and Astronomical. Above all, there was the Royal Society. At its meetings, conferences, soirées and conversaziones, scientists of all its branches congregated: Michael Faraday, Sir Humphrey Davy, Charles Babbage, Sir John Herschel, Dr Thomas Hodgkin, Sir Charles Wheatstone, Isambard Kingdom Brunel and many others, under the patronage of the Prince Consort.

Any study of Gideon Mantell opens up the social history of that age, the social structure of that time, the problems posed to dissidents within the community and how those not possessed of riches made their living. Besides his work on the Earth Sciences, Mantell was a medical practitioner and a skilful surgeon. The training of physicians, surgeons and apothecaries was beginning to become standardised. Mantell was involved in the development of his profession and wrote extensively in the medical journals of the day. His own ill-health – described in his journals – and the post-mortem report on his spine have been of interest to members of the medical profession. As a neurologist, I believe a new interpretation of his mental and physical problems is possible. However, I am also interested in the training and work of the profession in the nineteenth century, the search for knowledge in an age of burgeoning literacy, the development of scientific societies, and the debate and interaction of those determined to establish their own reputations in the scientific world.

The *Worthing Herald* of January 1928 described Mantell:

> He was not only an enthusiast in a science which has firmly established itself as one of the most important and fascinating branches of human study but he was an extremely able and plucky man who fought a stiff battle against a good many odds – ill-health, misfortune and, I think there is no doubt, jealousy on the part of other adventurers and theorists who were looking with keenly inquisitive eyes and penetrating intellect into the mysterious origins of life, human and animal.

His career could be summed up in the words he used when his museum was sold and set to be dispersed among the exhibits of the British Museum:

> Although my collection, the result of so much labour and toil, is lost to me forever, yet I do feel, and proudly feel, that I have not lived in vain. Although my country does not appreciate my labours, yet a time will come when they will be esteemed as I feel they deserve; and when I am no more, other minds will be taught by my example to pursue under difficulties and trials the way of truth and knowledge.

I believe a critique, rather than a straight biography, is needed to take the varied themes of his life and work and evaluate them, assessing and dissecting their significance.

Central to my study of Mantell is the library of the Sussex Archaeological Society at the Barbican in Lewes, opposite Castle Place where he lived, practised as a surgeon-apothecary, and started his museum collection, and in the town where he was born. I am indebted to John Bleach and the historians and staff working in the library, notably Esme Evans and Colin and Judith Brent. The bicentenary of Mantell's birth was marked by a symposium held nearby at the University of Sussex, Falmer, in 1990, when Professor Dennis R. Dean was the keynote speaker. The 150th anniversary of his death in 2002 was marked by a Mantell exhibition in the Sussex Archaeology building to which Anthony Brook, Colin Brent, and others contributed. With the considerable help of Christopher Milburn, overseen by the Society's archivist, Emma O'Connor, I have been able to make use of many of the illustrations used on that occasion.

Dennis R. Dean's primary research, on which his two books on Mantell are based, is derived from his examination of manuscripts and documents at the Alexander Turnbull Library, Wellington, New Zealand, via Mantell's elder son, Walter Baldock Durrant Mantell. Dean also recovered the fifth volume of Mantell's journal, held by his other son, Reginald Neville Mantell. Dean generously placed a typescript of the fifth volume together with the unabridged manuscript of his seminal work, *Gideon Mantell and the Discovery of Dinosaurs*, allowing ready access to all who wish to study the life and work of Mantell. I, personally, am very grateful to acknowledge this help, and have been in correspondence with Professor Dean.

I have also received personal help from Anthony Brook, a senior member of the Geological Society who has collated many of the articles on Mantell and contributed original material concerning Mantell, including aspects of his medical condition.

I am also grateful to Tim Lovell-Smith of the Alexander Turnbull Library, the East Sussex Record Office, the Lewes Library, and the Booth Museum of Natural History in Brighton. David Powell, historian of Tom Paine, advised me on the Headstrong Club. In London I wish to thank the archivists of the Wellcome Library (Lesley A. Hall), the Barts and the London NHS Trust (Justin Cavernelis-Frost) and the Worshipful Society of Apothecaries (Dee Cook), as well as the librarians of the Royal Society of Medicine, the Royal College of Surgeons, the Royal College of Physicians and the Royal College of Obstetrics and Gynaecology. My thanks to my publishers at Amberley, and in particular to Daniel Hill, Jasper Hadman, George Kalchev, Alan Sutton and Nicola Giles.

INFLUENCES

Dr Gideon Mantell (1790-1852) achieved fame as the discoverer of the Iguanodon and six other dinosaurs, but his additional involvement in the development of the emerging Earth Sciences – the stratification of rocks, the evolution of flora and fauna, archaeology and palaeontology – also deserves recognition. He settled uneasily among the other geologists of his time, lacking social status, needing to work at a profession to support his scientific endeavours, and considered by his detractors to be provincial. His obsessive personality and abilities clashed with many of his colleagues and especially with Professor Richard Owen who was determined to establish his particular supremacy by rewriting and re-interpreting the work of others in the same field. It would be fair and fully justified to claim that history has treated Mantell unfavourably and that a less biased assessment would have placed him among the foremost palaeontologists and scientific investigators of his day.

As a doctor (more correctly qualifying as a surgeon), his practice of medicine and midwifery, operative skill, therapeutic and other innovations, his writings particularly on local outbreaks of cholera and smallpox, and his diary descriptions of his own ailments, all provide remarkable insight into the practice of medicine away from the major hospitals.

The small, attractive, southern county town of Lewes, East Sussex, astride the nought degree line of longitude, where he was born, grew up and started his career, was slow to exhibit evidence of the Industrial Revolution but was well aware of world events. The townsfolk displayed and shared much of the intellectual, political, scientific, and religious fervour of the time; Mantell was to epitomise these aspirations. In so doing, his life, beliefs, struggles, ambitions, the controversies in which he was embroiled, and his accomplishments provide a remarkable personification of the social history of his day. Thus, before examining Gideon Mantell as an individual, we should give thought to Lewes and its environs and to his family background.

LEWES

Lewes can rightly claim a unique position in the history of our nation. At the Battle of Lewes (1264), forty-nine years after the signing of Magna Carta, Simon de Montfort, (brother-in-law of the king, and a disaffected French baron), together with merchants from London, defeated the army of Henry III. An insubstantial treaty, known as the Mise of Lewes, nonetheless enabled knights of the shires and burgesses of the boroughs to sit in parliament alongside the bishops and barons. But more importantly, the Battle of Lewes is seen as the birth of the English nation, bringing together Normans and Saxons as one people. Two years later in 1266, and again in 1334, 'murages' were instituted (taxes on the market to raise money) for building and repairs to the town walls, and in 1334 to erect the Barbican. These were timely, for in 1377, during the Hundred Years' War, the French raided Lewes capturing the Prior of Lewes, John de Charlin, who had led the resistance.

Within Lewes, religion ran with blood. The death of seventeen Protestant martyrs in the reign of Queen Mary is remembered by the traditional bonfires of the town, when six or more bonfire societies, complete with bands, banners, torches and fireworks, burn effigies of the Pope and others to whom the populace declare a dislike, and vats of burning tar on iron trolleys are ceremoniously rolled down the High Street ending up in the River Ouse. To many this is a hallowed tradition: to a few, including various Quakers and non-conformists through the ages, the ceremony is abhorrent.

Towns, such as Lewes, situated near the coast of southeast England, were fearful of disruption of trade by privateers operating in the Channel even in times of peace, and of the possibility of invasion from France during the Seven Years' War (1756-61), and again during the Napoleonic wars which followed the French Revolution of 1789. Fishing boats from the villages of the lower Ouse valley – Piddinghoe and Rodmell – would no longer be able to join the herring fleets off Yarmouth. The import and export of goods to Newhaven, at the mouth of the Ouse, and to Lewes itself would cease; and the building of barges in Lewes would come to a standstill.

The threat of invasion had a further significance. Families remembered the English Civil War when troops roamed the county. Families on both sides of the conflict were disgraced and impoverished. A new aristocracy arose and an old one recovered, seeking to become re-established. The effect of the Civil War on Sussex was uneven. A few of the gentry took up arms; others remained quiet, hoping to avoid embroilment in the conflict. Colonel Henry Morley of Glynde gathered troops, attacked Hastings, threw the curate into prison, and departed with a wagonload of arms. He then went to the west of the county where the war was more violent and joined Sir William Waller who recaptured the city of Chichester. Chichester was divided between, on one side, the Cathedral dignitaries and Sir Edward Ford of Uppmark who supported the king, and on the other, the townspeople led by William Cawley MP, who supported parliament. The first battle was to rescue the town from the royalists in the course of which the eastern suburb of St Pancras was burnt, destroying the local needlework industry. The town was then recaptured by royalists under Lord Hapton. Despite his role, Waller was imprisoned during the Commonwealth as he disagreed with the aims of Parliament, and was welcomed back after the restoration of the monarchy. Cawley, who had signed the king's death warrant, fled to France, and Sir Arthur Hasilrige, who used the war to amass huge church estates, was sent to the Tower.

The royalist Earl of Arundel had subscribed £12,000 to support the king. He abandoned Arundel Castle before the outbreak of hostilities in 1641, and fled the country the following year. The castle and town of Arundel were left with a weak royalist garrison who were besieged, relieved, then recaptured. During the Commonwealth, Morley, described as one of the ablest parliamentarians and promoted to Lieutenant of the Tower of London, formed a pressure group to counter the dictatorial powers of Cromwell before finding himself ousted from parliament. He obtained his pardon from Charles II at a cost of £1,000. At Petworth, Algernon Percy, the 10th Earl, held the younger children of Charles I in custody for the parliamentarians.

Mixed fortunes attended those living in East Sussex. Sir William Springett of Ringmer received fatal wounds. Sir Thomas Everfield and John Ashburnham had their estates sequestrated. Sir Thomas Evershed and Sir Herbert Lunsford were imprisoned after the Battle of Edgehill and transported to the Tower. Colonel Anthony Stapeley of Patcham, one of the men who had signed the king's death warrant, came to grief in the reign of Charles II. Sir Thomas Gages and Sir John Smith tried to avoid any confrontation but found themselves sequestered as recusants even though their delinquency could not be proved. Churches in the villages of Lullington, Arlington, Laughton and West Hoathly were desecrated. Lewes itself remained a firm stronghold for parliament with Captain Ambrose Trayton overseeing its defence. The only attempt to besiege the town resulted in a skirmish near Haywards Heath. The townsfolk showed no eagerness to become involved in the hostilities, despite strong Puritan religious beliefs. To quote Barbara Fleming (*Lewes: Two Thousand Years of History*, 1994): 'the conflict was essentially one between the lesser gentry and the clergy, who supported parliament, and the Crown who were supported in the main by the higher gentry.' Oliver Cromwell's successor, his ineffective son Richard, 'Tumbledown Dick', tried to establish a democratic form of government but fell foul of General Monck and the Army. The Rump Parliament's sterile and factional debate presaged an inevitable restoration of the monarchy. The Puritan era was overthrown. Yet to the consternation of the multitude, while the mob took revenge on Cromwell, disinterring his body and severing his head, the people of Lewes allowed Richard to avoid arrest and leave England by boat down the Ouse in July 1660 to a new life on the Continent. The escape was as much from his creditors as from the incoming king. The High Anglican Tories took revenge. With the Bill of Uniformity, 2,000 ministers were expelled from their livings and were forbidden to come within five miles of where they had preached. A further act, the Conventicle Act, meant that forty peaceable citizens of Lewes were prosecuted and had their property seized. The civic leaders of the town were ousted, thereby forcing Presbyterians and others – known as conventiclers – to worship in secrecy under threat of fine or imprisonment. Even so, they remained numerous, coherent and affluent, aided and eventually relieved through the mitigation of the more moderate Anglican, Sir Thomas Pelham (1653-1712), father of the Duke of Newcastle, who owned estates at Laughton, Halland and Bishopstone.

The Glorious Revolution, when William of Orange ousted James II, removed all possibility of a Catholic ascendancy, restored Lewes' Dissenters to civic influence, and gave them freedom to worship. During the next century the power of the Church of England continued to increase, though the majority of the populace became exceedingly lax in their religious practice. Religion centred round the pulpit with dry sermons aimed at reinforcing a state and religious orthodoxy as a way of life. The poorer clergy survived as curates and vicars, standing in for absentee benefices. However, better-placed rectors, rural deans and parsons were the protagonists of the established religion, rising in the social scale and living on equal terms with the gentry as never before. But their new cultural status often caused them to lose touch with their parishioners. Livings were in the gift of country gentlemen, worthy for endowment on their younger sons, and the bishops were chosen from relations of noblemen. Promotion within the church, as elsewhere, depended increasingly on Whig and Tory party patronage.

The Whig aristocracy came to the fore and achieved political dominance with the fall of the Tories after the death of Queen Anne and held power until the reign of George III. Their political henchmen were the dissident and commercial classes. Following the Hanoverian accession, Robert Walpole became the first prime minister and was succeeded by Henry Pelham of Stanmer (1743-54), who in turn was succeeded by his elder brother, Thomas Pelham-Hollies, Duke of Newcastle (1754-62). Newcastle had a house in Lewes, a hunting lodge at Bishopstone, and other estates in East Sussex. The brothers had very different temperaments: Henry was calm, almost phlegmatic, and after a fast-spending youth became conservative in his money usage; Thomas was excitable and readily agitated, always facing personal debt. Queen Caroline said of the Duke of Newcastle that he was such a mixture of fiddle-faddle and popularity that there was no making anything of him. Despite their apparent weaknesses, neither brother was ever accused of using office for personal gain or for being financially corrupt.

> He has no Private, Selfish Ends
> Esteems his honest, Steady Friends
>
> (Timothy Goose, 1741)

The two Sussex prime ministers are interred in the Pelham vault beneath the chancel of All Saints' Church, Laughton, twelve miles from Lewes. The Duke of Newcastle was one of the least successful premiers, but he held a grip on his party through firm management of constituencies in East Sussex, Nottinghamshire, Yorkshire and elsewhere. His extensive patronage meant that in his day he became notorious with the politically literate. He wanted men of ability but only if they were also docile. With such patronage there was the inevitable growth of corruption. Over half the higher clergy in Sussex were appointed by Newcastle. To cite a Sussex toast:

> You ne'er want while he's your head
> He gives you freedom, meat and bread...
> Your credit nor your stocks can't fall
> While Pelham represents you all.
>
> Then fill your glass. Full let it be
> Newcastle. Drink while you can see.
> With heart and voice all voters sing
> Long live Great Holles – Sussex King.

A more suitable quotation might be that of Thomas Paine: 'A long habit of not thinking a thing wrong, gives it a superficial appearance of being right.'

Under the Whigs, parliament maintained a balance between the Crown and people. The king could not dictate the policy of the cabinet but he could choose the prime minister and provide him with a 100 majority. He could occasionally veto parliamentary measures such as the attempt by the Younger Pitt to allow Catholic emancipation. In other respects, the aristocracy ruled with a certain degree of deference to public opinion and a respect for local and individual rights.

> This alliance between the spirit of aristocracy and the spirit of the popular rights, each taking the other entirely for granted, was native to the soil of England. It was satisfied by custom, sport and hospitality, deeply pledged in the punch-bowl, renewed on the hunting field and at the race-meeting. It was the natural offspring of a healthy society based … on the absence of very obvious economic oppression of class by class. The political spirit of the eighteenth century was based not on the equality but on the harmony of classes. (Trevelyan, *English Social History*)

With an emphasis, it could be said, on the sanctity of property. The divine authority of the king no longer existed, but the king was still powerful. As Henry Pelham was wont to assert,

> when parliament was against him, he might get his way by royal support; when the king disapproved, he sometimes got his way by relying on parliament; but against king and parliament he was powerless.

The Whig aristocracy were very comfortable in Lewes and well represented in parliament. They could even tolerate the occasional Tory within the town. Forming the local aristocracy, they divided their time between their houses in or near London, their estates in the Sussex countryside – especially during the hunting season – and their town houses in Lewes, improved with Georgian façades, for the winter season and the August race meeting. Lewes was the fashionable resort before the rise of Brighton. William Cobbett called it a town of clean windows and pretty faces. As the county town, it was also the centre for justice with its law courts and prison.

In this and other respects, Lewes was a vibrant oasis set in an area of poor soils and land. With the straightening of the lower Ouse and the improvement of its entrance at Newhaven, the river became navigable to barges and to some seafaring ships, enabling the town with its suburb of Cliffe to depend on a thriving import and export trade. Grain and wines could be imported; timber, bricks, lime, wool and iron goods were exported. Goods from Lewes were distributed over a wide area of the countryside. Fisherfolk brought their produce from Brighthelmstone along Juggs' Lane into Lewes. As a centre of commerce, it possessed banks, five breweries, tanneries, brick yards and kilns, chalk pits, iron foundries, wool and china merchants, paper mills, and the renowned Baxter print works. The sheep fair was in September, and cattle fairs were held on Whitsun and Michaelmas. The Whigs congregated at Newcastle House, the Tories at the Star Inn, which Thomas Sergison remodelled extravagantly, and the Headstrong Club, associated with the name of Thomas Paine, met at the White Hart. These and other sites were centres for vigorous political and religious debate.

The changes to the hardware trade, a result of the Industrial Revolution that began in 1770, meant nothing to Lewes until the coming of the railways. Along with the developments of canals elsewhere, improvements were made in the rivers to render them navigable. In 1733, the original outlet of the Ouse at Seaford had silted up and a new cut was made at Meeching (now called Newhaven). The river in the lower Ouse valley was straightened in 1791, and in the 1790s navigation north of Lewes was improved by the construction of cuts and locks. The roads in Sussex, passing through gault and heavy clay, were notoriously poor in winter, causing many accidents. One of the most famous was the overturning at Maresfield of the hearse conveying the body of Henry Pelham due to the bad condition of the road and the darkness of the evening. The six undertakers riding on top of the hearse were thrown to the ground, two of them receiving considerable injuries. Macadamised turnpike roads were not always appreciated as drovers objected to driving their animals over hard surfaces.

Changes in agriculture were more important. Enclosure of fields was a boon to the aristocracy.

> Inclosure of fields and commons I'm sure
> Can be called nothing better than robbing the poor.
> (Joseph Priestley)

Land was cheap. Large estates of 500 acres or more, with a few over 1,000 acres, dotted the Sussex landscape. The aristocracy built palatial mansions, and their 'broyles' and parkland provided hunting and shooting on a seasonal basis. Pigeon lofts, fish ponds, artificial rabbit warrens (coningerths) and deer parks attracted the sporting fraternity. The open land was used for hawking, fox-hunting and hare-coursing. For others, such as William Poole at Chailey, Lord Sheffield at Sheffield Park and Major-General Beatson at Mayfield, for whom the land was not just for display, the opportunity for agricultural innovation, which gathered pace in the last decade of the eighteenth century, led to improvements in drainage, drilling, rolling, double trenching, deep ploughing, manuring, and the use of lime, bone, and other additives. These innovations enhanced the value of the land, even achieving the utilisation of the heavy clay soils.

New practices in breeding and feeding stock meant that the wholesale slaughter of stock at the end of autumn was no longer necessary. The rising population in the latter part of the eighteenth century (overall a five to six-fold increase), better communications by rail and coastal traffic, and a general improvement in living standards augmented the demand for meat and dairy produce. The farmer responded with increased diversity in animal husbandry. Parallel advances included experiments in drainage, the substitution of cast iron for wood, stone and wrought iron, and the introduction of new machinery. Lord Egremont introduced the threshing machine in 1819. At the end of the eighteenth century, the iron plough replaced the wooden one with a lighter frame and a swing board to turn the soil over. Seed drills and horse drawn hoes were introduced by Jethro Tull.

Scientific farming in Sussex owed much to John Ellman (1753-1832), steward of the Trevor Estate at Glynde. He won prizes for cattle, some of his fat beasts weighing over 90 stone, but it was the development of Southdown sheep, cross-breeding the east and west Sussex varieties, for which he became famous. Short of leg, these beasts produced fine wool and meat. To achieve this, he introduced selective breeding, fast fattening, adjusting crop rotations to enable the feeding of stock on grains and hay. With the increasing use of fodder and a greater range of crops, cattle could be bred for purposes other than as draught animals. In fact a Sussex man, Albert Wadman of Firle, was the first farmer to send milk by rail to London. Crop rotations included clover and legumes and turnips, later to be replaced with swedes and mangle-worzels. So successful did Sussex farming become that a government survey of 1801 listed 350,000 sheep, 63,000 pigs, 60,000 cattle and 22,000 horses. Alongside the few notable innovators, a plethora of local agricultural and farming improvement associations sprang up throughout the kingdom. By 1810, there were over seventy such societies.

The spreading development of enclosures achieved four objects for the landowner: more profit, better care for the animals, easier management, and the alternation of use between arable and husbandry. Even so, continuing extravagance meant that even some of the nobility lived close to bankruptcy. The smaller farmers, though paying high rents, were generally successful; but the cottager changed from a labourer with land to one without land. However, there was an increase in the volume and regularity of employment with the need for hedging, new roads, farm houses and outbuildings, and heavier crops requiring more hands at harvest time. The more fortunate were compensated by the availability of allotments and cottage gardens. For those not successful, the increasing birth rate meant more rural unemployment, poverty, and a rise in the need for poor relief. In this respect, many landowners were able to control who lived in their villages, manipulating the number of cottages and the size of the workforce on their estates in order to minimise pauperism in their locality. Thus the Sussex peasant remained poor, scraping a living and at the mercy of the stewards of the larger estates. The male workforce was partly seasonal and forever fearful of the introduction of newly developed machinery. Few females or children were employed.

The traditional alternatives to agriculture remained strong, the making of bricks and pottery was largely seasonal, but smuggling was as successful as ever. Smuggling began centuries earlier with 'owling' – the export of wool avoiding tariffs – later involving the import of brandy, tea, silks and other goods. All sections of the community were involved. Pelham's chaplain at Stanmer was wounded in a fight with revenue men and the Vicar of Rottingdean owned a fast grey mare and often acted as look-out when there was a run-in at the Gap. Even those seeking re-election to parliament often had to tread carefully for fear of upsetting their voters. In 1741, electors told the Duke of Richmond they would vote for his candidate if he released a smuggler from gaol, otherwise they would vote against him. Lewes town possessed the cellars etc. which could have been used, but it is probable with a successful legal economy that they were less interested in becoming involved. From the time Prince Charles[1] escaped to France from Shoreham after arriving in disguise at Brighthelmstone, religious dissidents and others had been smuggled into exile. Catholic priests, Jacobites, French aristocrats, Huguenots, and even spies formed part of the human cargo at various times. With the French Revolution this aspect of smuggling gained prominence. The Scarlet Pimpernel was no myth. Some of the nobility escaped by packet-boat to Newhaven, or used smugglers to land them along the coast and then gained sanctuary in the great houses of the Sussex gentry. The Comte de Llally, for example, sheltered as guest of the first Lord Sheffield at Sheffield Park.

There was an association with Lewes in Mantell's time. A Mr Penderell, who took over the White Horse Inn in St Mary's Street, was a linear descendent of Richard Penderell who concealed Charles II in the oak and still received an annuity from the crown for the services of his ancestor. He changed the inn sign to The Royal Oak.

So far I have stressed the peregrinations and political impact of the gentry on the fortunes of Lewes town and its environs. The town itself was justly praised. In the 1720s, Defoe declared it 'full of gentlemen of good families and fortune'. It was later described as a garden town of garden walls and trees, built on a slope with elaborate gardens at the foot of the castle and high up the opposite slope on Cliffe hill. William Cobbett visited the town in 1822, declaring:

> the town itself is a model of solidarity and neatness. The buildings are substantiated to the very outskirts, the pavements good and complete, the shops nice and clean, the people well-dressed and the girls reasonably pretty.

By contrast, he described Brighthelmstone, the future Brighton, as 'a poor fishing town, old built, and on the very edge of the sea, which has been very unkind to this town by its continual encroachments.'

THE MANTELL FAMILY

Once the Catholic succession was put to rest and the early sectarian quarrels lost meaning, the clergy loosened their hold in the towns and the state religion fell into decline. The growth of literacy enabled religion to become increasingly laicised. Nonconformist evangelical movements took root with enthusiasm. The greatest and most justly famous of the movements was Methodism, spurred on by the revivalist preaching of the Wesleys and Whitefield. Methodism was so named because it called for greater discipline and regularity in the conduct of life. In Lewes, the artisan class which had embraced Whig politics also predominantly and enthusiastically took to the 'dissident' religious sects. There were Quakers, active as brewers, ship-builders, wool-merchants, timber-merchants and iron-mongers. The Wesleys did not come to Lewes but Selina, Countess of Huntingdon (1709-91) established the Independents in Lewes. Other sects and chapels differed in their degree of Calvinism and belief in predestination. It was the peoples' religion, not fed to them by the priest as if they were children. They, as individuals and not the state, were the church. They possessed a reverence for literature and an interest in public conduct, service and ethics based on the Bible, sermon reading and private prayer. They were evangelical, bright in rectitude, unselfishness, and humanity. Being desirous to help the downtrodden, the miserable, the ignorant and the debauched – those who had fallen from society – Methodism spread through the trading and professional classes. Its devotees were not withdrawn from the business of life, but strove to dedicate it to God. The work ethic developed in parallel to an intensely earnest devotion to their faith – a faith undiluted by being compelled to pay parish rates and maintain the church fabric.

Such was the ethos of the Mantell family as exhibited by Gideon Mantell's father. Previous biographers have described Thomas Mantell (1750-1807) as an unlettered shoemaker or bootmaker, but to do so is to underestimate Thomas's standing as an artisan in the town. From his will, we understand that he employed as many as twenty-three journeymen and apprentices. Even if he just made bespoke footwear and riding boots with spurs attached, he would have made a good living. If we describe him correctly as a cordwainer,[2] then he might also have made other leather merchandise like riding breeches, gloves, saddle bags, box-cloth for coaches and aprons.

There were three tanneries unsurprisingly situated at various points on the extreme periphery of the town. Tanneries were among the smelliest and most unhealthy places conceivable. The skins were treated in vast vats of lime, ammonia, acid, and dung to clean and macerate the flesh, and then impregnated with tannin by lying in a solution of powdered oak bark before being dressed, coloured, or softened in yet further noxious mixtures. Thomas would need various types of leather for his trade and would naturally have overseen how leather for his shop was procured, cured, tanned, and primed. The tanneries made Lewes a centre for trade in leather goods, with considerable competition, thereby suggesting that people came from a distance to purchase more deluxe items. By 1820, thirteen years after Thomas Mantell's death, there were forty master shoemakers in the town. A humbler tradesman might have lived above the shop. His house was in St Mary's Lane (now Station Road) with his business in Fisher Street. The sizeable premises in Fisher Street were adjacent to the brewhouse yard and appurtenances, and were purchased by Thomas Mantell from Thomas Sergison. All this might explain how he came to build the first Wesleyan chapel in town and owned several properties in St John Street and St Mary's Lane. He may have had little formal education but almost certainly he read voraciously, even if his reading was confined to Sundays. 'Just knows and only knows his Bible true. The truth the brilliant Frenchman never knew' (after Cowper[3]).

For three generations the Mantells were friendly with the Lee family who started and owned the first newspaper in Lewes. Gideon recorded his appreciation of the Lewes Library Society in Albion Street with its 10,000 volumes containing standard works of history, science and general literature which he was able to use as a small boy. One can infer that Thomas Mantell's handwriting was probably poor, as despite reading as widely as possible, he probably had little need to write. This theory is backed up as his signature, when he was ill and possibly had diminished vision two years before he died, was noticeably infirm. But though he read as widely as possible, he probably had little need to write.

Gideon Mantell was very conscious of his 'humble' origins, coming from an artisan class, to be a fashionable medical practitioner rather than having descended as a scion of the aristocracy. But as Dr Johnson declared: 'An English tradesman is a new species of gentleman; wealth has dressed itself in liberal behaviour and given itself airs commonly passing muster for gentility.' Gideon's brother, Thomas Austen Mantell (approximately fifteen years older than Gideon), was able to become a sheriff's officer, local politician, and auctioneer, despite the fact that dissenters were barred from academic life or any sort of public service. They were supposed to make their living from trade, industry or property. The explanation is probably quite simple. The Marriage Act of 1753 required that marriage rites could only take place legally in the Church of England, except for Quakers and Jews. Thus it was necessary to attend the church occasionally and in the case of magisterial or municipal office to have been confirmed and be a communicant. Towns such as Lewes could be relatively lax in their discrimination in the selection of those seeking public office, allowing some leeway to those respected within the community. Vestiges of the discrimination against Catholics and nonconformists have continued down the centuries. At Bexhill, East Sussex, until the 1960s, members of the Church of England were buried in the cemetery at the top of the hill, others further down. Similarly in Little Horwood, Buckinghamshire, in the 1940s, couples married in chapel, in the presence of the registrar, and presented themselves in the parish church the following Sunday in order that their marriage was acknowledged by the village.

For three generations, Thomas Mantell's ancestors had held the office of headborough.[4] The brother of John Mantell, his son and grandson were all headboroughs. Six generations later, Thomas is recorded as a cordwainer in Lewes. Thomas was the fourth son, given the same name as the third son who died in infancy as was the custom. The eldest was the Reverend George Mantell (1757-1832), Gideon's uncle. Thomas, like his father, married an Austen. Sarah, his wife, came from Kent and is described as the ruling spirit of the household, but Thomas was always noted for his shrewdness, integrity, and Whig principles.

There were nine children: the eldest, Sarah, died in infancy; Thomas Austen; Samuel Augustus, who became a butcher and inn-keeper, had six children and lived in a poorer part of town; Mary; Gideon Algernon – Gideon was a judge of Israel called into action by an angel to defeat the Midianites; Algernon Sidney,[5] a republican hero; Joshua, who was born delicate with a hunchback and was initially thought unlikely to survive; and two other sisters, Jemima and Kezia. All received an early training in religion from pious parents; Gideon's retentive memory enabled him when young to repeat a large part of the Bible by heart. When Thomas

Mantell senior died in 1807, he was buried in St Michael's Church. A neat iron railing was placed on the grave of the father and mother, and a yew tree, a few rose trees and spring flowers were planted within the area. The incumbent Reverend Peter Guerin Crofts, better known on the hunting field than in pursuit of his ecclesiastical duties, was pleased to take great offence at this, and compelled Gideon to take up the yew tree, allowing the railing to remain as a great favour – so much for 'priestly pride, arrogance and utter want of taste and every finer feeling. Talk of catholic intolerance what can exceed this protestant overbearing spirit?' Thus deriding this exhibition of sacerdotalism, Mantell erected a tablet in St Michael's Church in memory of his father:

> Though humble was the lot to thee assign'd
> The sterling virtues of an upright mind
> In thee, beloved parent, purely shone,
> And made content and happiness thy own
> And though no sculpture mark thy lowly grave,
> Nor yew nor cypress o'er thy relics wave,
> The virtuous will respect the hallowed spot,
> When prouder names than thine shall be forgot,
> Oh! Fain would he who in these feeble lays
> Attempts a father's and a good man's praise
> Follow the bright example thou hast giv'n,
> And humbly trace thy footsteps up to heaven

This was almost certainly written by Gideon, whose diary contained a considerable amount of his versifying.

Gideon and his father both stuck firmly to their Whig and dissenting principles, but they sought to establish their noble genealogy. The name Mantell was Norman-French in origin, coming to England with the Norman Conquest. There are still descendents of Mauntell or de Mantell around the town of Abbeville. A William de Mantell accompanied Richard the Lionheart on the crusades. In 1541, a John Mantell was among a group – Lord Dacre, John Frouds, and George Royden – poaching from Sir Nicholas Pelham's deer park at Halland, in the course of which they murdered a gamekeeper. All four were subsequently executed and their estates forfeited to the Crown. John's son, Walter, then joined Sir Thomas Wyatt's rebellion in 1554 aimed at preventing Queen Mary's union with Philip of Spain. Walter, his nephew, and Wyatt were all executed the same year. In Charles II's time, members of the family in Lewes were fined or imprisoned for being 'conventiclers'.

DISSENSION THROUGH THE AGES: JOHN WILKES AND THOMAS PAINE

> Give me the liberty to know, to utter, and to argue freely according to conscience, above all liberties.

These words from the peroration of John Milton's *Areopagitica* of 1644, directed against the proposed licensing of the printed word have inspired liberals and dissidents down the ages. Licensing, he argued, was used by those whom the Presbyterian government most detested, viz. the Papacy and the Inquisition, while Moses, David, St Paul and the Fathers enjoin freedom in the pursuit of learning. Throughout the eighteenth century, the battle for liberty was fought on many fronts. Those who did not conform to the mores of the land, the kings, the establishment, and the state church, and attacked corruption in high places, were branded along with Roman Catholics as unpatriotic and a potential threat to the stability of the kingdom. Hence the insistence on outward forms of adherence before they could be regarded as responsible citizens.

Royalists were strongest where the economic and social changes were least felt, but from the beginnings of the eighteenth century they entered a time of change. Troubles with France erupted in India and then in the Americas with the battles for Canada and the removal of France and Spain from America. The settlements in America included the Pilgrim Fathers who had escaped religious persecution in the previous century, nobles who opened up Virginia, and convicts sent to work the plantations before the slave trade took over the cotton fields. Attempts to maintain the colony without representation under British law were to lead to the American War of Independence (1775-83). From 1735 to 1738, a great awakening of Methodism in America began with the visits of the Wesleys and George Whitefield. There was no similar pilgrimage of bishops of the Church of England. In 1750 the distant government imposed the Iron Act whereby pig iron could be exported from the States, but no iron products made. These had to be imported and taxed. The Townsend Revenue Act of 1767 imposed tax duties on tea, glass, paint, oil, lead, and paper. In March 1770, the first bloodshed of the American Revolution occurred as British guards at the Boston Customs House opened fire on a crowd, killing five. Among the issues involved were the presence of troops, competition between soldiers and civilians for jobs, and the shooting of a Boston boy by a customs official. Better remembered was the Boston Tea Party of 1773, resulting in the destruction of 342 chests of 'dutied' tea by working men disguised as Mohawks, after other ports had refused to let the tea ships enter. Lord North failed to avert the Declaration of Independence of the North American colonies in 1776 and to defeat them in the subsequent War of Independence, 1776-83.

In Britain, the Hanoverian succession brought stability of government, church, and law, but remained essentially Germanic. George II, despite being king for twenty years, loathed the country and all things British. This was reciprocated by the populace who derided with contempt his self-indulgence with mistresses and extended visits to Hanover. The man who tested the freedom of the press to its limits was John Wilkes (1727-97), editor of the *North Briton*. Descriptions of him are far from flattering. He was exceptionally ugly with a pronounced squint, but possessed of an insinuating charm such that it took him 'only half an hour to talk away his face'. Elected an MP for Aylesbury in 1757, he attacked the king and the ministry, was imprisoned, released, then thrown into the Tower but saved when a judge ruled that this violated his parliamentary privilege. He escaped to France and was expelled from the House of Commons on several occasions. As a result he came to be seen as a champion of liberty and an upholder of press freedom. Thousands protested outside the King's Bench prison where he was held, and in the Goodman's Field Massacre several demonstrators were shot.

Wilkesite clubs were among the spate of reformist, radical and counter-revolutionary societies of the late Georgian eras. According to Josiah Wedgwood, Wilkes and Liberty were tagged together like 'Hobgoblins and Darkness'. His visit to Lewes in 1770 was greeted with peals of bells and applauding crowds shouting 'Wilkes and Liberty'. Tom Paine almost certainly witnessed the event. In 1774 Wilkes became Lord Mayor of London and in the same year finally regained his seat in parliament where he remained until his retirement. Invited to play a game of cards he replied, 'Do not ask me, for I am so ignorant that I cannot tell the difference between a king and a knave.' In answer to a prediction that he would die of the pox or on a gibbet, he said: 'That depends, my lord, whether I embrace your mistress or your principles'. Wilkes epitomised the general disaffection with the monarchy and corrupt government. Even Dr Johnson, who was later to receive a state pension himself, defined pensions thus: 'In England it is generally understood to mean pay given to a state hireling for treason to his country.'

George III, known as 'farmer George', restored some respect for the monarchy as the first English Hanoverian king, and attempted to deal with corruption in parliament with a rapid succession of unfortunate prime ministers. Britain was saved from invasion during the Seven Years War (1756-63) by the strength of its maritime fleet. Religious evangelism gained force with the otherwise orthodox philanthropist, William Wilberforce (1759-1833). He urged the aristocracy to give a moral lead to the poor and to promote welfare schemes. He believed, quoting the words of the seventeenth-century James Naylor, that God made all men of one mould and one blood to dwell on the face of the earth. However, he is best known for his fight against slavery, which was strongly supported by the Nonconformist section of the community. The French had abolished slavery in their colonies in 1794. Wilberforce began his campaign when elected to parliament in 1788, but slavery was not abolished until 1807. (Wilberforce had a tenuous connection with Sussex; passing near Bramber, he was reminded that this was his parliamentary constituency!)

Political discussion in Lewes was invigorated by the presence of Thomas Paine at the Headstrong Club which met at the White Hart Inn. Almost certainly, Gideon's father, Thomas Mantell, between the ages of eighteen and twenty-four, was either part of the club or very aware of its debates. Thomas Paine (then spelt Pain) (1737-1809) was born in Thetford. His mother, Frances, *née* Cocke, was aged forty, an Anglican and the daughter of a Thetford lawyer and town clerk. His father, Joseph, was a Quaker aged twenty-nine, who though elected a freeman of the town, was considered somewhat beneath his wife, earning £30 a year as a smallholder and a maker of corsets (stays). However, Joseph's earnings enabled Thomas to attend grammar school where he learnt history, mathematics, and science, but not Latin or Greek. Thomas was baptised and later confirmed in the Church of England but continued to attend both denominations. He left school at thirteen and was then apprenticed to his father for four years before escaping to London, walking forty-eight miles. Seeing a notice 'for gentlemen sailors and able-bodied landsmen to try their fortunes and serve their king and country fighting the French on board the *Terrible* a privateer, Captain, William Death!', he was on his way to the dock when his father retrieved him, preventing his enlistment. The privateer was sunk by a French boat in the channel with only seventeen of its crew surviving.

Thomas made several attempts to make a living on land before enrolling, this time successfully, on the *King of Prussia*, under the command of Captain Mendez. Once at sea, the *King of Prussia* captured eight enemy vessels before returning to London the following year. He then left the ship and tried to find work as a staymaker, eventually setting up on his own in Sandwich. There he married Mary Lambert whose father had been a customs officer or exciseman. When Mary died, presumably in childbirth, Thomas returned to Thetford. Unsuccessful as a staymaker, he decided to follow his father-in-law's profession and took an exam which enabled him to enrol as an exciseman. Excise duties had to be paid on imports of tea, chocolate, coffee, tobacco and alcohol. His work took him to Grantham with responsibility to examine brewer's casks and he was later allocated the more dangerous duty of patrolling a section of the Lincolnshire coastline on horseback to intercept smugglers. He escaped being harmed by the smugglers only to be dismissed for the misdemeanour of stamping a consignment of goods which he admitted he had not examined, accepting the shopkeeper's word instead of making a personal assessment of the goods. He was fired, but two years later in 1767 he managed to get the decision reversed.

Once reinstated, Paine was posted to Lewes. Sussex especially was a notoriously difficult and dangerous assignment for an exciseman. It was an important but also a precarious post, as smuggling was almost a recognised profession in Sussex and Officers of Excise where not popular. Smuggling was indeed rife, with well-armed gangs landing gin and other contraband and few revenue officers of any description to prevent their incursions. His supervisor was the manager of the White Hart, Thomas Scrace, who oversaw nine excise officers. Paine was billeted at Bull House, owned by the Ollive family who sold tobacco, snuff, cheese, butter, and other groceries (but not tea). Samuel Ollive, the head of the family, was a respected town elder, worshipping both as a member of the Church of England and a Quaker. He belonged to the Society of Twelve, a governing body of the local elite, and served the community as a constable and as a member of the St Michael's Vestry aiding the poor of the parish. Paine (T. Pain) also served on the Society of Twelve from 1770-73. Two years after Samuel's death, Paine married his daughter, Elizabeth, and in effect became involved in the business.

Paine's six years in Lewes enabled him to blossom in his personality and to mature, articulate, and amplify his opinions. He was regarded as popular, friendly, outgoing, and possessing charm. Curious to relate, he had an enviable reputation as a cricketer, a skater on ice, and an expert on the bowling-green. William Lee, editor of the *Sussex Weekly Advertiser*, called him a shrewd and sensible fellow who enjoyed a depth of political knowledge; Lee published many of Paine's comments under various noms de plumes in his paper. Paine attended meetings of the Headstrong Club. It was essentially a heavy drinking club given to boisterous debate and disputation. There was no known membership. William Verrall was the landlord of the White Hart and Henry, his brother, owned a coffee house. Thus it is likely that the more ebullient goings-on became public knowledge. Many others – including, presumably, Rickman, Lee, Button, and Thomas Mantell – steeped in the traditions of Whig principles and dissent, were sympathetic to the nascent radical who could entertain them with witty sallies and fierce opinions. From time to time, the participants at the club were prone to add verses to their 'Headstrong Book, or book of obstinacy'. One such poem by Clio Rickman to a farmer's dog, related an event when a farmer near Shoreham had voted contrary to the wishes of the local grandee. The borough magistrates could not convict him of any crime but sought revenge on his dog, finding it guilty of the death of a hare, and sentencing it to hanging.

> That this dog did then and there
> Pursue and take, and kill a hare
> Which reason was, or some such thing
> Against our Sovereign Lord the King
>
> So had the dog not chased the hare
> She'd never had drowned – that's clear
> Thus logic, rhetoric, and wit,
> So nicely did the matter fit,
> That Porter – though unheard – was cast,
> And in a halter breathed his last.
> The justices then adjourned to dine
> And wet their logic up with wine.

As the leading disputant of the club, Paine was elected General of the Headstrong War with an anonymous and prescient poem:

> Immortal Paine, while mighty reasons jar
> We crown thee General of the Headstrong War
> Thy logic vanquished error, and thy mind
> No bounds but those of right and truth confined
> Thy soul of fire must sure ascend the sky
> Immortal PAINE, thy flame can never die
> For men like the their names must ever save
> From the black edicts of the tyrant grave.

Paine's fall from grace came through his agreeing to place before parliament the case for an improvement of the salaries of the Officers of Excise. He printed 4,000 copies of his pamphlet setting out their grievances. He then went to London distributing copies to Members of Parliament and prominent persons including Oliver Goldsmith. However well argued the pamphlet was, it was easy for the authorities to dismiss the claims, as excisemen were treated with contempt by all sections of society. Paine was duly relieved of his duties for failing to remain at his post.

Paine then sought new worlds. He left his wife claiming the marriage had not been consummated and in London called on Benjamin Franklin, the respected unofficial ambassador for the American Colonies. Franklin wrote Paine a letter of introduction to his son-in-law in America (which was more of a testimonial than a warm recommendation).

> The bearer, Mr Thomas Paine, is very well recommended to me as an ingenious, worthy young man. He goes to Pennsylvania with a view to settling there. I request that you give him your best advice and countenance as he is quite a stranger there. If you can put him in a way of obtaining employment as a clerk, or assistant tutor in a school, or assistant surveyor, (of all of which I think him very capable,) so that he may procure a subsistence at least, till he can make acquaintance and obtain a knowledge of the country, you will do well, and much oblige your affectionate father.

He was found a job as a tutor but almost immediately went into journalism.

In absentia, Paine's activities still had an indirect impact on the dissident population of Lewes. Arriving in America in 1774 he became editor of the *Pennsylvania Magazine*, publishing in 1776 a pamphlet, *Common Sense*, which attacked slavery and advocated full independence for the American Colonies. During the Revolutionary War he served in Washington's army and continued his political writings, eventually acting as secretary to a committee on foreign affairs. A further pamphlet in 1780, *Public Good*, reiterated the case for a federal union as opposed to the Virginia plan. He returned to England in 1787, staying in London with Clio Rickman, whom he knew from Lewes, and replied to Burke's *Reflections on the Revolution in France* with *The Rights of Man*, published in two parts in 1791-92. It urged the overthrow of the British monarchy, and, in the second part, the abolition of all hereditary titles. A foretaste of his opinion is recorded from his time in Lewes. In response to a remark from Harry Verrall that 'Frederick of Prussia was an excellent king, having "so much of the devil in him,"' Paine replied, 'kings might well be disposed of.'

The tract prompted the government, headed by William Pitt the Younger, to introduce a law against seditious publications, and Paine, who had fled to France, was convicted of sedition in his absence and was outlawed. In Paris, the revolutionary assembly made Paine a citizen of France and a deputy. In the volatile atmosphere of the revolution, his proposal that the King of France should be given asylum in the USA led to his imprisonment where he wrote his last pamphlet of note, *The Age of Reason* (1794-96), a stark critique of accepted religious beliefs and practices. In a bald statement he decreed that 'all national institutions of churches, whether Jewish, Christian, or Turkish, appear to me no other than human inventions set up to terrify and enslave mankind, and monopolise power and profit.'

Following the reign of terror, James Monroe secured his release, but Paine's return to the USA led to ostracism by his former friends. As an infidel when he died, he could not be buried on consecrated ground. Cobbett secured his bones ten years later and brought them to England, but burial was not allowed and his bones then disappeared. Only his skull and right hand were later recovered. If the men of Lewes sympathised with his views on the monarchy, American independence, and the French Revolution, his denunciation of religion in the *Age of Reason* turned them to anger.

Paine may have influenced opinion within Lewes, but throughout the kingdom, following the storming of the Bastille in July 1789, and the French Revolution in its wake, campaigners in various parts of Britain contrasted the religious freedom claimed in France and the position of Dissenters at home. The second attempt to repeal the Test Acts had been defeated by just twenty votes. Richard Price, a Dissenting Minister, called for equal representation and the repeal of the Test Acts, hailing the new regime in America and the revolution in France while demanding that the rights of mankind be restored. Edmund Burke, who had previously supported many liberal causes, formed part of a vocal minority, presenting the revolution as the wrecking of all institutions, the vandalism of culture, and the replacement of humanity. To read, he warned, is to lay oneself open to contagion!

THE FORMATIVE YEARS

Gideon Mantell was born on 3 February 1790. The storming of the Bastille on 12 July 1789 had marked the beginning of the French Revolution. Before starting school he had learnt to read and recite passages from the Bible, taught by other members of the family. Then aged six he attended a dame school nearby. The site of this school is uncertain. Dean places it in Fish (Fisher) street, while Cadbury claims that it was among the cottages in St Mary's Lane.[6] A year later he moved to John Button's academy for boys in Cliffe, across the road from St Thomas's church in Cliffe high street.

Although Methodists and others were denied entrance to universities or to grammar schools, Mantell's education could not be described as frugal, i.e. economical and parsimonious. The Dissenting Academies supplied, at moderate cost, a good education where their own people received higher and secondary education. They were the real centres of science and intellect, conducted on more modern lines than the grammar schools or even Oxford and Cambridge. Living languages and science held a place along with the classics. Button was a prominent Baptist dissenter, an openly radical Whig remembered as a 'gentleman whose political sentiments were so accordant with those of Mr Mantell's father that he was known to be on the Government's black list.'

Button emphasised: grammar, rhetoric, composition, and penmanship. Arithmetic, geography, classical and British history were given preference over Latin and Greek. The navigation teacher was enlisted fresh from a man-of-war. He insisted upon daily oral recitations with the better students performing in public. Through these lessons in elocution his pupils were expected to achieve verbal fluency and expunge 'vulgar and provincial' diction.

The seven-year-old Mantell's success with these public performances was reported by friends of the family in the local newspaper – either the

Sussex Weekly Advertiser or the *Lewes Journal*. In the assembly room at the Star Inn he presented an extract from Homer's *Iliad*: The Parting of Hector and Andromache. In an evening of elocution at the Star Assembly room he spoke of an imaginary 'Trip to Paris' that included a particularly repulsive dinner with snails. At the age of eleven he wrote a book for his sister, Jemima, featuring eight coloured drawings which he had done of sites including Lewes Castle and the priory. That same year he wrote a verse of fifty-six lines, 'To His Uncle' describing a characteristically tumultuous general election in Lewes and the victory celebrations that followed. Out of school he made full use of the town's well-equipped lending library. The Lewes Library Society had been founded by a surgeon, John Ridge, in 1785 to acquire 'quite serious' books and embracing books in every science and containing works too expensive for individual purchase. 'His breast conceiv'd the gen'rous thought'. The library was maintained by sizeable subscriptions from among the dissenting and radical section of the community who would vote on which books to obtain. The stated object was to enlighten people's minds with a much wider range of subjects than that available at Oxford or Cambridge. Originally there were just twenty-

Lewes Castle, from Mantell's garden.

eight members and 1,000 volumes, but by the turn of the century there were over ninety members and at least 10,000 volumes.

In 1803 he left Mr Button's academy to live with that same uncle, the Reverend George Mantell, a Congregational minister, Pastor of the Independent Congregation of the Upper Meeting House in Westbury, Wiltshire where he was taught alongside the Rev. Mantell's son, a year older than Gideon. In January 1804 they all moved to Swindon when the Reverend Mantell was appointed minister to the Newport Street Independent Chapel and head of a Dissenting Academy for boys. Gideon's departure from Lewes raises many questions. It is unlikely that any of his siblings left Lewes to be educated. The answer seems to have been political. Not only was Mr Button on the government's blacklist, or even it was alleged on Pitt's own blacklist; but, according to Gideon Mantell, writing some time later, 'My poor father suffered greatly in fortune during the war with revolutionary France for his Whig principles.' Button was described as a dangerous demagogue, an exuberant philosophical radical who read the *Examiner* and admired Charles James Fox. When his boarders drained away, removed by their Tory parents, he declared his loyalty to the constitution as shaped by the Glorious Revolution.

There was also Pitt's gagging act, The Treasonable Practices Bill, which forbade the meeting together of fifty or more persons without the consent of a magistrate; any meeting could be dispersed and the speakers arrested. This was a significant ratcheting up of Pitt's repressive response to discontent with severe penalties imposed on those who attacked the constitution or gave support to the nation's enemies. His Attorney General declared in 1794 that it was high treason for any man to agitate for the establishment of 'representative government' – the direct contrary of the government which is established here. The townsfolk of Lewes particularly came under the spotlight because of their known Whig principles and their past association with Thomas Paine who had left England for France. Questions were on everyone's lips: would the revolution come to England, will the monarchy end, should the common man have greater democracy?

Local anger of every kind was aroused. Lord Sheffield[7], who masterminded the passage of the French nobility and other émigrés through Sussex en route to London, chaired a meeting to assist those 'who by unexampled barbarity are driven on our shores'. With the Terror at its height, the meeting resolved to preserve Liberty and Property against Republicans and Levellers. The resultant hysteria took various forms. Effigies of Paine were gibbeted at Seaford and burnt elsewhere. 'Marked men' were named and many took flight. James Drowley, a lay preacher, and Paul Dunvan fled to the United States. John Button and William Elphict were listed as dangerous demagogues but stayed firm. With objections to the Gagging

Acts occurring in other parts of the country, a county meeting in November 1795 debated the acts with Arthur Lee, Thomas Rickman, Henry Browne and Thomas Mantell among those objecting. In December that year the local MP Thomas Kemp presented a petition to parliament against the Gagging Acts signed by 300 residents. The fiefdom of Newcastle was no longer in line with the beliefs of government.

The change in direction of Gideon's education at the age of thirteen was almost certainly the result of a family conference. From the poem to his uncle, it is evident that the Rev. Mantell was aware both of Gideon's intellect and the political atmosphere in Lewes, and kept in touch with his ancestral home. He probably saw Gideon as a companion and stimulus for his own son and may have believed that he could provide a more classical education that that of Button's academy. Gideon's father was obviously concerned for his future. As a younger son Gideon would not inherit his father's business; his ability and wide interests meant that like his uncle he would bridge the gap in social status between trade and profession. But as a dissident a formal university education was not open to him. Gideon himself may already have resolved to follow a medical career, rather than law or the ministry. In all likelihood the decision was arrived at rapidly and without hesitation Gideon set off with his uncle to Wiltshire.

The move from Lewes was beneficial both for Gideon's future education in a calmer atmosphere and it enabled Thomas to loosen a potentially dangerous association with John Button's actions. The political and religious principles on which Gideon was fed would not be weakened, but there would be less risk of public agitation. John Button was perhaps more subtle in escaping retribution. His elder son, John Viney Button, very much a contemporary of Gideon, and known as the Laureate of the Library Society, was elected on 22 October 1807 by the constables of Lewes, Arthur Lee and William Kennard, to an exhibition or scholarship to the University of Cambridge[8], funded by the will of the late Rev. George Steer. He eventually took orders as an Anglican priest. Gideon kept up a correspondence with his former master who advised him to write poetry for his own amusement: 'No workman is prouder of his work than a poet and you will always refer to what you have written with pleasure.' The other son, William Button, remained a radical Methodist.

Gideon by this time had grown as a tall, slim boy. The Reverend George Mantell was remembered as very tall, with gigantic shoulders and amazing strength. Like Gideon, the young George Mantell, also took up a career in medicine. Gideon's scholastic career, short though it was, was distinguished by uncommon perseverance and quickness in his studies and a good knowledge of French and Latin. While at Swindon, Gideon apparently had a schoolboy crush. This event would have passed unnoticed except that in

December 1826 he received a letter from his uncle informing Gideon of the death of a lady, formerly Miss Strange, 'to whom when a schoolboy, I was much attached'. The following year, on a visit to Swindon he went to the churchyard 'where so many of my old friends had taken up their abode, and strolled through the 'long walk' where in happier hour and circumstances, I had been accustomed to walk with my "*first love*"'.

What free time a child had would naturally be spent wandering off, alone or with friends to discover his surroundings. Almost certainly he would walk for hours and miles; if lucky it is possible that he would be able to ride over the countryside. Certainly Gideon learnt to ride well before it was an essential for his medical practice. Lewes derives its name from the Anglo-Saxon word for hill, 'hlaew'. It is built on a hilly spur sloping down towards the Ouse, crossing to the market site and hamlet of Cliffe and ascending a steep escarpment on the other side. To the west is the site of the Battle of Lewes and Mount Harry, with the Downs, in places wooded, extending to Brighton and to Ditchling beacon and beyond. To the east is the primitive fortification of Mount Caburn with its tumuli, and beyond the 'bare-backed' Downs extend towards Eastbourne at Beachy Head. He would have been aware of Saxon terracing (lyncheting) of fields on the Downs. The Napoleonic wars meant that wheat could not be imported so, for the first time since the Saxons, the chalk downs were ploughed and sown with grain. Mantell would possibly have been unaware of the Saxon inhabitations on Itford hill or at Black patch. But he would have known about the erosive force of the sea. Parts of Shoreham and a large part of old Brightelmstone had disappeared into the Channel. Longshore drift had blocked the estuary at Seaford, necessitating a new cut at Newhaven and diverted the Arun at Shoreham. The river Cuckmere had its cuspate delta. At the Seven Sisters he would not only have observed the crumbling cliffs but in the eighteenth and nineteenth centuries there were vertical pinnacles projecting from the middle segment of the cliffs between Belle Tout and Headledge. These outliers of the main cliff line have been referred to as stacks, but since they developed two-thirds the way up the cliff face, they should more correctly be called turrets. Known as the seven Charlies they disintegrated and collapsed one by one, the last falling in 1875.

The need to use his powers of observation came as an eight year old boy from reading 'improving literature'. The book in question was Anna Letitia Barbauld's *Eyes or No Eyes*, or *the Art of Seeing*, suggesting that rather than being bored when out walking one should be curious and note the plants and other objects, including a probable Roman camp! M. A. Lower when listing Mantell among 'The Worthies of Sussex' described how Mantell collected fossils when he was twelve or thirteen:

> While yet a mere youth he was walking one summer evening with a friend on the banks of a stream communicating with the Ouse, when his observant eye rested upon an object that had rolled down from a marly bank which at that particular spot overhangs the stream. He dragged it from the water and examined it with great attention. "What is it?" was the natural inquiry of his friend. "I think, Warren," he replied, "that it is what they call a fossil. I have seen something like it in an old volume of the *Gentleman's Magazine*." The 'curiosity' which proved to be a fine specimen of the ammonite, was borne home in triumph by the two friends and from that moment young Mantell became a geologist.

Warren was Stewart Warren Lee, son of the local newspaper editor. The *Gentleman's Magazine* stimulated interest in antiquities, the architecture of earlier periods and a sense of heritage. Fossils were commonly known before the development of the science of geology by such dialect terms as pundits (echinoderms), snakestones (ammonites), and devil's fingers (belemnites). They were called the wreckage of former lives that had turned to stone.

His other 'juvenile explorations', enhanced by studying geological phenomena in Albion Street's Lewes Library Society Building, included drawings of capitals among the ruins of Lewes Priory and collecting encaustic tiles from the site. William Lee, the newspaper publisher introduced Mantell to Rev. James Douglas (1753-1819) of Preston, near Brighton, a local archaeologist. With his encouragement Gideon collected items from various tumuli on Mt Caburn and Oxsettle Botton. One barrow, illustrated in Horsfield's *History and Antiquities of Lewes and its Vicinity*, which was excavated by Mantell held an urn packed with calcinated bones; and also yielded a brass ring, a bone armille and beads of amber, jet and 'green porcelain'. Douglas was a fellow of the Society of Antiquities and had published a *Nenia Britannica*, or a *Sepulchral History of Great Britain* between 1786 and 1793 – a multi-issue directory of burial rites and customs of the Celts, Romans and Saxons. He personally supervised the opening of several hundred downland barrows in the vicinity of Lewes in the decade from 1770, and remained Gideon's mentor until his death. Douglas was born in 1756 in Holland and enlisted as an officer in the Austrian Army. As Captain Douglas he made his first excavations of tumuli but in 1780 was ordained from Peterhouse College, Cambridge. There is some doubt as to when he first entered the priesthood. According to Lower (*Worthies of Sussex*) he was tempted into the clerical profession for the sake of 'learned leisure'. However, Rosemary Sweet (*Antiquaries: The Discovery of the Past in Eighteenth Century Britain*) states that he served as a military chaplain when they excavated the Chatham Lines, and his interest in antiquities stemmed from that experience. While in Brighton he became chaplain to the Prince of Wales. The Rev. James Douglas and Gideon Mantell acted with ebullient enthusiasm, opening the tumuli and documenting their treasures, in their urgency to visit newly discovered sites and the spontaneity of their euphoria at handling the relics of earlier cultures. Their finds were recorded and removed, and often lost. The Curwens, father and son, as the next generation of archaeologists, would have condemned their methods; and in their turn we would condemn them for lacking a truly scientific approach and not ensuring the preservation of their findings.

At that time, and certainly in the eyes of the young Gideon Mantell, Douglas' approach to the collection of data, making comparisons, drawing inferences and testing hypotheses against the evidence, provided the bench mark for that generation. Antiquities, he claimed, were the alembic (distillate) rather than the dregs of history. He stipulated that when deductions were made they should be founded upon a scrupulous comparison of fact. In his *Nenia Britannica* he analysed the excavated barrows, making a distinction between people who worked with metals and those who used only flints as weapons. He used the differences in the contents of the burial site to achieve greater certainty in dating the tumuli. For example, from first-hand observation and systemic excavation, comparing the different types of tumuli, he drew up a sophisticated typology which enabled him to claim that the small conical barrows found in the South East of England were of Saxon origin.

The same cleric was aware of the work of William Smith (1799) a surveyor and civil engineer, who in the course of building canals had developed the first geological map of England and circulated an account of his method for tracing the strata by noting the organised fossils embedded within each layer. In a letter of 7 April 1813 the Rev. James Douglas proffered Gideon with sound advice: 'the great desideratum of geology is now fixed on Strata in which the bones of quadrupeds are found, they determine the last great revolution of our planet, antecedent to our present order of created life.' (Anthony Brook, *Gideon Mantell; Memento Mori* vol. 3) Douglas inculcated into Gideon the yet-to-be-accepted belief that the earth's history consisted of a series of distinct epochs separated from one another by episodes of violent changes.

When Gideon was at Swindon, his curiosity was strongly excited by the petrified 'rams horns' and 'oak' so abundant in the solid masses of stone in the neighbouring quarries. The quarrymen would regale him with wonderful stories of petrified beaks and crocodiles. Although describing himself as too young to understand the nature of the fossils, curiosity tempted him to pick up a few specimens and begin a collection. His uncle explained that they had been left by the biblical deluge. On his return to Lewes friends encouraged him to continue collecting fossils, showing him a mammoth's tooth brought from America. Later, in 1804, following a collapse of 6,000 tons of chalk at Robert Hillman's pit off South Street, Gideon found his first fossil fish. His collection was further enhanced when in 1809 he went on a walking tour of the Isle of Sheppey with his friend, John Tilney.

On 15 January 1805, Gideon finished his scholastic training tutored by his uncle, the Reverend George Mantell, and returned to Lewes. Gideon like his cousin George wished to take up a career in medicine, and this was arranged with the help of a local Whig leader and attorney, Mr William Cooper, who was impressed by Gideon's readiness of perception and strictness of integrity. On 4 February 1805, one day after his fifteenth birthday, he enrolled as an apprentice surgeon. Surgery and, in some places, medicine, were treated as crafts under the Statute of Artificers (1563) enabling the aspiring 'artisan' over a period of up to seven years to learn his craft under a responsible master, building his character and providing what amounted to a higher education. For surgery, boys were expected to have received a grammar school education in the classics until the age of 12 or 15 and most (62 per cent) were the relatives of established practitioners, or of clergy or lawyers of the same social set.

The start of the nineteenth century saw the development of the medical profession divided between Physic, Surgeon and Apothecary. In the eighteenth century the term medical practitioner referred to anyone whose living derived largely from treating the sick. Many clergy and 'practical' farmers undertook diverse aspects of treatment. Fully qualified practitioners were rare; and in the provinces medical practitioners who were not affiliated to the College of Physicians were supposed, by an act of 1511, to be licensed by the bishop of the diocese acting in conjunction with expert assessors; but by the seventeenth century these stipulations had fallen into decline.

Within the City of London, the College of Physicians, the Barber-Surgeons and later the Surgeons College were more concerned with encroachment upon their own privileges and spheres of practice than with the control of the numerous quacks and charlatans who flourished within the metropolis. Surgeons severed their links with the Company of Barbers in 1743 when the Company of Surgeons was established. The Royal College in London was founded in 1800 and became the Royal College of Surgeons of England in 1843. Apothecaries were originally members of the Company of Grocers but formed their own society in 1615, supposedly dispensing prescriptions written by a physician. The education for a medical degree at Oxford or Cambridge, open only to communicants of the Church of England, was lengthy and expensive with an emphasis on classics, Hippocrates and Galen. They dominated the College of Physicians of London and their superior education was supposed to embrace the whole of surgery and pharmacy.

Edinburgh, Glasgow and Aberdeen had more reasonable tuition fees, a respected medical faculty, clinical lectures and extra-mural teaching, but at the Scottish colleges few bothered to take medical degrees, having served an apprenticeship. MD degrees could be obtained by thesis on the Continent (e.g. Leyden) or by payment in Scotland from 1750. In London, St Bartholomew's Hospital in 1820 had only three medical students and several hundred surgical students. Within England consulting physicians and pure surgeons were few and far between; the majority mixed physic, surgery, pharmacy and midwifery. The term general practitioner was first used in 1809 and survived in respect of the surgeon-apothecary until 1860.

Mr James Moore, the surgeon and medical practitioner, to whom he was apprenticed, was a much respected figure in the locality. Moore had spent his own apprenticeship in Lewes, and after studying under John Hunter in London, had returned to enter into practice. He experimented with lithotomy for stone, and published in 1794 an *Essay on the Medicinal Qualities of Steel and its mode of operation on the Human Habit*. Moore would have appealed to Gideon not only as a collector of fossils and minerals but also on account of his acknowledged reputation as a botanist.

Indenture as an apprentice was a solemn occasion, swearing an oath of servitude for five years and promising to keep his master's secrets; not to waste, gamble, haunt taverns, commit fornication, go absent without permission or marry during that time. Thomas Mantell agreed to pay Moore £50 with a further £50 in June the following year, and continue to provide clothing, laundry and dinners for his son. Moore, in turn, undertook to teach Gideon his trade while furnishing him food, lodging and medical care. All three participants signed the document. It was Thomas Mantell's signature on this occasion which led biographers to question the extent of his education.

Apprenticeship, a proud fraternity, was set up by the Statute of Artificers (1563) enabling aspiring artisans to learn their craft under a responsible master, building their character and providing what amounted to a higher education. It is clear that Mantell took full advantage of this opportunity studying anatomy, blood vessels as well as bones, reading widely while Moore, with his interest in natural history and the newly found excitement as to what fossils might represent, was prepared for Gideon to devote some time in their collection and examination.

Gideon moved into 63 High Street, sleeping behind the surgery so that he could answer the night bell, rouse his master and accompany him as required. His first duties included sweeping and cleaning the surgery, delivering medicines and collecting the used phials. He was taught how to make pills and other pharmaceuticals. With no method of evaluation, many apprentices were mistreated and did not progress beyond a menial stage. Much depended on the teaching provided by the master. After the first year he wrote out the bills, kept the accounts, bled patients and extracted teeth. Being tall and looking old for his age, he visited the poor with slight ailments, and treated simple fractures and minor surgical injuries in parish patients. He assisted Moore in a series of post-mortem examinations, reports of which were then sent to the Director-General, Dr Rollo. Later he would treat most fractures, 'debride' and close wounds and assist childbirth. Along with developing the skills of a surgeon and apothecary, he studied avidly gaining a systematic knowledge of anatomy which he wrote up and illustrated in his notebooks. With the result that by the time he had finished his apprenticeship he could claim

> to have comprehended many things which I ought not to have been required to perform. My knowledge of anatomy was equal to that possessed by most of the hospital's students of my own age, for the habit of copying drawings and making extracts of standard works had familiarised me with the elements and nomenclature of the science and given me a good theoretical acquaintance with its details, so that the lectures on osteology were readily comprehended, and by the frequent dissection of animals during my apprenticeship I had acquired a dexterity in the use of the scalpel which greatly facilitated the practice of human anatomy. In consequence I soon gained the expected notice, and ultimately the friendship, of my excellent and eminent teacher.

Thomas Mantell died in 1807. He had five daughters and four sons, and though his will delineated over fifteen properties which he had owned, his effects would have had to be divided between his two elder sons and their families, and his widow with responsibility for the other children. As a part of her share of the will, the widow sold the Wesleyan chapel. This meant that the fees required for Gideon to go to London to complete his medical training may have had to be specially arranged (according to Cadbury) with the assistance of friends. There is no record of his time in London. All we know is that he took with him to London the best of his collection of fossils in order to make the acquaintance of the leading geologists in the city. A student decided his own curriculum, attending lectures as he wished, besides walking the wards. If he preferred he could choose to attend lectures at several different hospitals or private schools. Students paid no lecture fees to a hospital, but could purchase admission tickets to as many individual courses as they wished to attend. Each lecturer sold tickets for his own courses. At the end of a course a certificate of attendance might be granted to those who had completed it. Certificates of 'hospital practice' were also issued to students who had attended regularly in the wards. These certificates elaborately framed could be exhibited in lieu of a diploma or proper qualification. Surgeons at Bartholomew's Hospital were giving private lectures in the 1720s, and such was their popularity that by 1788 Barts had built its own lecture theatre. In 1822, the surgeon Abernethy persuaded the Governors to establish a Medical School. At St Bartholomew's there were rarely more than three medical students, while the surgical students, amounting to several hundred, never entered a medical ward.

That Mantell gained the notice of Abernethy is not surprising, but to have achieved his friendship until he achieved fame would have been unusual. John Abernethy (1764-1831), Surgeon at St Bartholomew's and Professor of Anatomy and Surgery at the Royal College of Surgeons, epitomised the tradition of the arrogant surgeon throughout the ages by putting down his pupils and being rude and abrupt with his patients. Even so he was a gifted teacher, clear, succinct, impressive and fascinating. 'We never left his lecture room without thinking him the prince of physiologists and ourselves only just one degree below him.'

A later student (Dr Thomas Pettigrew), writing in 1840, described him thus:

> The lecture room was the grand theatre upon which Mr Abernethy displayed; there, indeed, he shone eccentric like a comet's blaze and there he would indulge his disposition and propensities to an extent which occasioned the pupils frequently to regard it as an exhibition, and call it an Abernethy at home. His mode of entering the lecture room was often irresistibly droll – his hands buried deep in his breeches-pockets, his body bent slouchingly forward, blowing or whistling, his eyes twinkling beneath their arches, and his lower jaw thrown considerably beneath the upper. Then he would cast himself into a chair, swing one of his legs over the arm of it, and commence his lecture in the most outré manner. The abruptness, however, never failed to command silence and rivet attention.
>
> Having described in the most bloodthirsty terms a soldier wounded in battle, he would enter upon an admirable discourse on the nature of gunshot wounds, their peculiar character, the course they followed. Then he would contrast the improvements that had been made by the moderns

over the practice of the ancients; and by the relation of often ludicrous and always interesting anecdotes, fix the subject on the minds of his pupils in the most indelible manner.

On 18 April 1811, he certified that Mr Gideon Mantell 'hath diligently attended the anatomy lectures, those on the theory and practice of surgery and dissection from 1 October to the present time'. Anatomy was regarded as the queen of medical sciences. The practice of dissection of cadavers by medical students had been introduced by William Hunter (1718-83). The Murder Act 1752 stipulated that the corpses of executed murderers could be used for dissection but by the nineteenth century the shortage of bodies had attracted the lucrative trade of body snatchers.

Mantell relates in his *Memoirs of the Life of a Country Surgeon* (1848) that students were sadly restricted in their studies by the scarcity and high price of cadavers for dissection; £6-10 being the usual price for a body. Abernethy probably acquired cadavers for his students, as had William Hunter, through some sort of private arrangement. It is alleged that Sir Astley Cooper was a great dissector and a voracious client of the 'resurrection men'. The history of the procurement of cadavers is more complicated. The teaching of anatomy in London was confined at one time to lectures and demonstrations at the Barber-Surgeon's Hall on the bodies of criminals executed at Tyburn (J. Dobson, *The Anatomising of Criminals*, 1951). From Queen Elizabeth's time these demonstrations were held four times a year and lasted three days. Private anatomies could also be held at the hall by persons given permission from the company and later from the college. It sometimes happened that an overzealous teacher omitted the formality of obtaining permission. William Cheselden, as recorded in the Annals of the Barber-Surgeons of 1724, obtained the bodies of malefactors and privately dissected them in his own house lecturing during the course of the dissections. Later it became customary for the Master of the Company of Surgeons and afterwards the Conservator of the Museum of the Royal Colleges of Surgeons, to make at least a preliminary or token dissection of such bodies of criminals that had been directed by the judge to be 'anatomised'.

In the Charters (fortieth George III, 22 March 1800, and third George IV, 13 February 1822) it is stated 'that the said College shall, and by these Presents they are required to purchase or provide a proper Room, House, or Building, with suitable Conveniences, within 400 yards, at the farthest, from the usual Place of Execution for the County of Middlesex, or the City of London and the suburbs thereof'. As Lincoln's Inn Fields, the new site of the College of Surgeons, was outside this limit, a warehouse in Castle Street, West Smithfield, was first used for this purpose. Only after 1832 was there a requirement for those practicing anatomy to obtain a licence from the Home Secretary. Apart from dressing ulcers, there is no clear indication of what practical experience a student might get in surgery.

Mantell passed his oral exams at the Royal College of Surgeons – an hour's questioning on anatomy and surgery – before the President, Sir Everard Home, and was enrolled with the diploma of Master of the Royal College of Surgeons by Mr Edmond Balfour. On the 23rd he also received the certificate of attendance in midwifery from the Lying-in Charity for Married Women at Their Own Habitation (instituted in 1757). V. Squire MD, the charity's physician solemnly recalled how Gideon had diligently attended his lectures in the theory and practice of midwifery and the diseases of women and children; during which time the hardworking student had had many opportunities of delivering women in difficult and dangerous cases as well as natural labours, and with the happiest success so as to fully qualify him to engage in practice. These certificates including statements of attendance on wards were often displayed on the walls of their surgeries.

The use of the word 'diligent' by both lecturers suggests that this was the standard mantra of the day reminiscent of the medical school joke of the chief attending a dinner given by twenty or thirty younger people and replying to a toast to his health with the words: 'In thanking you I express my surprise, some of you I know well, others I think I can recognise, pray please explain yourselves.' To which came the reply that 'each of us has a testimonial from you in our pockets saying that we were the best house-physician you ever had.'

At this stage he hoped to obtain an assistantship under a London practitioner, but this was not to be and he returned to Lewes, voting for the first time in the election of October 1812. His old mentor, Dr James Moore, offered him a small share of the practice. He accepted although it did not appear likely to produce more than £50 per annum. The practice depended in great measure on midwifery, the fee for attendance rarely exceeding 10/*6d*. Three parishes within Lewes held contracts with the practice at £20 each. Surgical and obstetric cases in the parishes were individually priced, receiving extra payment at low rates.

Thus in 1811, at the age of twenty-one, newly qualified, Gideon Mantell faced the world, fired by a restless energy with an ambition to match. He was a handsome youth with expressive features and dark hair and eyes. From Button's lessons in elocution he had developed a pleasing voice with clear diction and the stage presence of an actor. He was not above self-congratulation. He could extract teeth better than his mentor, James Moore. And, to adapt a later comment, 'You must know that being tall and slender, my fingers are proportionately slight and flexible, and I have therefore a very light hand and delicacy of touch; this will account for my giving less pain in my manual operations.' He could claim serious scholastic achievements. As a versifier he matched the skill of his contemporary, John Viney Button. He had recited in public, been quoted in the local papers, and working alongside his elder cousin had been a cause of stimulation to him. He had spent hours in self-education, reading in libraries, collecting and learning about fossils and almost certainly deserving the adjective diligent.

His father, the Rev. George Mantell and John Button had all been valuable role-models inculcating religion and a deep sympathy, understanding and desire to help the underdog: whether he be the farm labourer kept in a state of positive ignorance and slavery, those involved in revolutions abroad or the victims of the Peterloo massacre. From the Rev. James Douglas he had learnt about the landscape with its tumuli and the rites of the dead of former centuries, and from his own efforts and enthusiasm he had become a collector and investigator of fossils and the strata in which they were found. From the start of his practice he reveals himself to his biographers, both in his works and in his diary, as a very human personality with abundant energy. His politics and behaviour reveal a moral fervour, keenly aware of his 'humble' origins. He soon evolved into a polymath fired by a craving to become immortal in the realms of science. All obstacles were irritating frustrations which he bore with the weight of a dominant personality.

The first years of the nineteenth century were a period of change, scientific and revolutionary and of open communication. This was the Age of Enlightenment. News could be thrilling and unexpected with the growth of literature, of political pamphlets passed from person to person and the advent of vicious cartoons. People learnt of the plight of those fleeing the French Revolution, the reign of terror, and the Napoleonic wars at sea and on land. They learnt of the pamphlets and escapades of Thomas Paine in France and latterly in America. The manner of communication could appear sensational, thus Wordsworth when crossing the distal part of the Duddon estuary from Holmgate to Ulverston, was met by a passing horseman:

> A variegated crowd
> Of vehicles and travellers, horse and foal
> Wading beneath the conduct of their guide
> In loose procession through the shallow stream
> Of inland waters – the foremost of the band
> As he approached, no salutation given
> In the familiar language of the day
> Cried 'Robespierre is dead!'

For eight years he continued in partnership and through incessant labour increased the returns from £250 to £750 per annum by dint of attending from two to three hundred cases of midwifery annually. He was frequently up six or seven nights in succession – an occasional hour's sleep in his clothes being the only repose. Also on returning to Lewes Gideon eventually replaced Moore as military surgeon at Ringmer, a position he soon came to despise in view of the frequent ferocious floggings with the hospital seldom free from recruits suffering from the infliction of this horrible and bestial punishment

In 1813 he went to London to meet James Parkinson and learn about his compendium of fossils, *Organic Remains of a Former World.*

> He kindly showed and explained to me the principal objects in his cabinet and pointed out every source of information on fossil remains – a department of natural knowledge at that time but little cultivated in England. Mr Parkinson warmly encouraged my attempts to elucidate the nature of the strata and organic remains of my native county, Sussex, a district which was then supposed to be destitute of geological interest and he revised my drawings and favoured me with his remarks on my subjects treated in my first work on the Fossils of the South Downs.

Much to his chagrin he was unable to afford the cost of buying the work. However, he had the good fortune to attend and minister to a patient in London, a relative of John Tilney, with whom he had been on a walking tour round the Isle of Sheppey to collect fossils. The patient, George Edward Woodhouse, was a prosperous gentleman who owned a linen business. His father James Woodhouse, had been a poetical shoemaker and author of several books. In payment for his services, Mr Woodhouse gave as a gift all three volumes of Parkinson's opus: 'I beg the acceptance of this work as a small token of my gratitude and esteem, and as a trifling acknowledgment of the peculiar professional exertions and kind attention which I have experienced from you during my illness'.

This was both a most generous gift and for Mantell an important acquisition. Mantell's unremitting professional attention was not eventually successful and Mr Woodhouse died; but Gideon had in the meantime formed an attachment to the patient's eldest daughter. Mary was an attractive, dark-featured lass. Her portrait shows her with masses of dark curls piled high, and large regarding eyes, her face set off by a fashionable off the shoulder dress. She shared Gideon's interest in fossils and gave him gifts of corals and other curiosities. They were married in May 1816.

He was determined to make himself known within Lewes as an active member of the Lewes Philosophical Society which met in the Star Inn, along with G. M. Horsfield and John Dudeney. Henry Browne Jnr lectured on horology and Gideon's brother, Joshua Mantell, on silver, performing experiments with a blowpipe and other apparatus. Members of the defunct Philosophical Society later became prominent in the Lewes Mechanics Institute, founded in November 1825 with the aim of the widest possible dissemination of knowledge. Within twelve months it had recruited over 200 members; mostly tradesmen, and housed 500 books in purpose-built premises adjacent to the theatre. Gideon Mantell lectured on arsenic, Dudeney on astronomy, Woollgar on mechanics, Horsfield on pneumatics, Godlee on electricity and steam power and Henry Browne, jnr. on radiation of heat.

In 1815 James Moore bought No. 3 Castle Place but probably did not intend to reside in the building. Who lived there initially is unclear. The town records state that it was 'now in the occupation of James Moore and Gideon Mantell or one of them'. The uncertainty was resolved in March 1816 when James Moore moved to All Saints Parish and Gideon rented Castle Place from where he practised. By May of that year Moore's health began to deteriorate and on 25 March 1818 the partnership was dissolved. Gideon had only to pay Moore £81 per annum for sevem years and £40 annual rent. In June that year he hired his own assistant and bought the premises outright for £725.

In May 1819 he bought the adjoining house, No. 2 off Amon Wilds for £600. He was anxious to have the most elegant and sophisticated town house and employed a local mason and architect to combine the two houses with entrances to the new drawing room and the front bedroom from the passage ways of No. 3. This was performed with considerable ostentation. The end houses, Nos 1 and 4, were later pulled down but have been rebuilt. Both the architect Amon Wilds and Gideon Mantell were enthralled by the prospect of employing the ammonite motif evolved by George Dance who had blazoned it to enhance the façade of Boydell's Gallery in Pall Mall in 1789. It was augmented by pilasters, the volutes in the capitals of which were designed in the form of ammonites – geological fossils – consisting of whorled chambered shells resembling the horn of Jupiter Ammon in shape. Amon Wilds and his son Amon Henry Wilds[9] must have known of the existence of this order and, besides Gideon's natural attraction to a fossil façade, were probably attracted by the allusion to their own Christian names.

Dr Gordon Hake (1809-95, *Memories of Eighty Years*) written forty years after Mantell's death was unimpressed with the flash of his surroundings.

Castle Place, High Street, Lewes. Mantell's residence, 1821-33.

> The house was fronted by ionic columns and pilasters with an ammonite for the volutes. His gig and groom were models as they waited at his door. His coat of arms embraced the visitor as it shone in the fanlight and whispered greatness within.

Of the man he had a better opinion:

> Lewes is a feeble, antiquated presentiment of civilisation ... but it contained Gideon Mantell, the great geologist who searching Tilgate Forest became the discoverer of the Iguanodon. He struggled for fame by his researches within the chalk strata, and for his livelihood by his practice as a surgeon and apothecary, in which he had a fair amount of success, no doubt due to his great abilities. He was tall, graciously graceful, and flexible, a naturalist realising his own lordship of creation.

M. A. Lower describes him as tall, thin, and of a rather light complexion, with a keen penetrating eye, a clear musical enunciation and of an aristocratic bearing.

The interior would gradually come to display all the trappings of success. The second floor overlooking the street was its drawing room which doubled in size with the combination of the two houses in 1819. His fossil collection expanded from a cabinet to a museum. On display were a collection of Napoleonic medals, two allegorical paintings, an elaborate gilded mirror, a very elegant French clock, a likeness of a portrait of Oliver Cromwell, two Wedgwood vases, Bartolozzi engravings[10], two heads by John Hamilton Mortimer, and a marble-topped table with Mantell and Woodhouse heraldry painted side by side within a single shield. The Mantell arms were on a silver shield compartmentalised by a cross with each quadrant containing a black martlet. A frontally depicted stag's head surmounted the shield. The scene was completed with the chaise and groom standing outside.

At a time when every trade and profession could be recognised by their dress, physicians, surgeons and general practitioners vied with Beau Brummel in their appearance and were ostentatious in their choice of horses, coaches and coachmen. Their status supposedly added to the reassurance they gave their patients. Likewise Mantell, in his personal demeanour, was intent on keeping up a most professional appearance suitable for a fashionable surgeon. His portrait shows him impeccably dressed in black, silk hat, Brummel haircut, sideburns, jacket and waistcoat, high pointed collar, goffered shirt, cotton gloves, plain or striped trousers and black Wellington boots. In winter he might wear a cloak or greatcoat and a cravat fastened with a diamond amethyst brooch, which on one occasion he lost on the way to Glynde. However, a labourer found it and returned it intact. For his operations and other procedures he probably protected his clothes with a large leather apron.

Spokes in his biography of Mantell states that he drank little, and indeed in his diaries there are no references to drinking apart from one when he took brandy to relieve a headache. In the eighteenth century, as at the Headstrong Club, heavy drinking was the norm and the Younger Pitt drank six bottles a day. Mantell's second elder brother Samuel ran a pub so it is unlikely that Mantell's family were teetotal, but Mantell as an apprentice until the age of twenty-one was forbidden to frequent taverns and when in London probably did not have the money, other than paying for his keep and training, to indulge himself, quite apart from his ambition which provided a strong discipline. The beginning of the nineteenth century was one of progressive refinement with pride and haughtiness of soul. Politeness and good breeding would open doors. The swearing of the gin-palaces had been replaced by the conversation of the tea table. Self-restraint and appreciation of art were part of the professional bearing.

This outward show did nothing to detract from his intense application to his work. As a medical practitioner he had a gruelling workload with forty to fifty patients to see or visit each day and 200-300 midwifery cases to attend to each year. There were epidemics of cholera, typhus, typhoid and smallpox and countless serious accidents needing surgical attention. With patients in childbirth he might be up six to seven nights in a row. One statistic suggests the success of his practice; Dr Bisset Hawkins in his *Elements of Medical Statistics* (1829) quotes from the report of a practice in a healthy provincial town (Lewes) published by Mr Mantell (1828). In fifteen years there were 2,410 cases with only two deaths of mothers. Dr Hawkins remarks 'this document forms a remarkable contrast to the register of the lying-in hospitals of great cities. It proceeds from a gentleman whose name is familiar to the friends of science.' The mortality rate at the City of London Lying-in Hospitals was 1 in 70, in Edinburgh 1 in 100, and at Dublin 1 in 223. He became locally famous for his midnight industry as he read and wrote in the quiet of his study. He maintained serious interests in botany, zoology, astronomy and history and the study of antiquities. He continued to sketch and draw throughout his life but always as an adjunct to science.

He kept abreast of developments in medicine seeking the advice of Dr John Armstrong of the London Fever Institute on the treatment of typhus and a Mr Pearson and Sir Astley Cooper in respect of gangrene. He was not averse to obtaining a second opinion especially regarding his more influential patients. He obtained the opinions of both Sir Henry Halford, President of the Royal College of Physicians, and of Sir Benjamin Brodie,

President of the Royal College of Surgeons. They both approved his programme of treatment and could offer no alternatives. In his *Memoirs of the Life of a Country Surgeon*, he complains that he was required to amputate and perform other operations without assistance. However, he did work with a senior surgeon, Thomas Hodson, and wrote a fulsome obituary when he died. Together they removed a fungus haematodes (a cancerous growth) from the side of a poor man who had had it for years, operated on a tumour of the breast, and on an inguinal hernia. He was called in consultation by Mr Hodson for a gunshot wound to the wrist causing profuse haemorrhage. He assisted him in the amputation of the thigh of a man with a diseased knee joint; and Mr Hodson in turn assisted Mantell in deftly amputating a mangled forearm.

Lewes had six or more general practitioners, five of whom were officially qualified as surgeons (as opposed to physicians) when the twenty-one-year-old Mantell entered into partnership with James Moore. The terms general practitioner and surgeon were virtually synonymous until the 1880s. Such a practitioner

> presides at birth and is sought in death, he knows of the skeleton which is hidden in most cupboards, he knows the tragedies which blast a life, and the minor discords which embitter it, he sees human virtue and human frailty, joy and sorrow, life in its seaminess and life in its excellence. Well for him if he can preserve a genial toleration for the frailties of mankind, and the spirit of charity. (*The Lancet*, 1919)

Likewise in 1826 the Sussex County Hospital, Brighton had nine physicians, twenty-nine surgeons and four dentists.

In the eighteenth century, disease described the body's reaction to disequilibrium arising from the imbalance of the bodily humours as the result of external influences. One form of treatment, particularly for mental illness, was diversion; thus Robert Burton in the *Anatomy of Melancholy* (1621) stated that there was 'no better physick for a melancholy man than a change of aire and variety of places. To travel abroad and see fashion'. Visiting the spas and taking the waters, on the continent if that could be afforded or increasingly within the United Kingdom, became the fashion. The health appeal of resorts rested on their capacity to combine the traditional magico-religious elements of therapy with new empirical scientific ones. In 1841 Dr A. B. Granville constructed a map of England showing seventy principal mineral springs and thirty-six principal sea-bathing places; the latter coming into prominence with Dr Russell's advice on the value of sea-bathing. Dr Russell proved, to the eighteenth century's satisfaction that the sea was not only sublime but also healthful.

Dr Russell FRS obtained his doctor's degree from Leyden for a thesis on epilepsy. He believed the ocean to be nature's defence against all bodily corruption and putrefaction. Called by the historian Jules Michelet, 'the inventor of the sea', Russell postulated in his suitably lurid *Dissertation on the Use of Seawater on Diseases of the Glands* that sea water was a cure for scrofula, consumption and glandular fever. But all people and patients were advised to bathe – preferably early in the morning – and drink sea water at various temperatures for their health. He performed wide-ranging experiments with concoctions from crab's eyes to cuttlefish bones and they were rubbed with fresh sea-weed and sniffed fumes of boiling 'stromboli' – a fuel dug from the seashore. He also promoted chalybite at St Anne's Well in Hove – this was a siderite in a hydrothermal mineral vein containing iron carbonate in crystalline form. George IV's surgeon, Sake Dean Mohamed, likewise advocated vapour baths. Russell's successor in Brighton, Dr Anthony Relham, who took over most of his practice published *A Short History of Brighthelston with Remarks on its Air, and Analysis of its Waters, particularly of an uncommon mineral one, long discovered though but lately used by Anthony Relham, MD, Fellow of the Royal College of Physicians in Ireland*, extolling the virtues of the soil and weather; and the purity of sea air to convalesce from illnesses:

And all, with ails in heart and lungs,
In liver or in spine,
Rush coastward to be cured like tongues,
By dipping into brine.

(Brighton, *A Comic Sketch*, 1830)

More people took baths to ease ailments, if not for cleanliness. Improvements to the roads and the coming of the railways lessened the number of accidents to travellers. However, advances in sanitation only came about towards the end of the nineteenth century.

*

Mantell retained a close relationship with James Moore whose health continued to deteriorate. In December, within a few months of retiring, he had a most severe fit of apoplexy. Mantell reports:

> I was with him when it came on and immediately made a large orifice in a vein and took away 40 oz of blood. I then cupped him, had leeches applied, etc. He had convulsions repeatedly for three days successively. Dr Blair and Mr Hodson thought his case hopeless. I left him one

> morning (the third after the fit) apparently dying. I wanted to bleed him again but Mr Hodson would not consent. He thought it useless and that he should be left to his fate. I was obliged to go into the country to visit a patient some distance from Lewes; upon my return in the evening, I found Mr Moore in the same state as I had left him, pulse slow and full; breathing stertorous; convulsions every quarter hour, he was entirely senseless. I resolved to bleed him again as the dernier resort. I took from him 20 oz; this induced to me not to stop the bleeding till his pulse faltered; he amended further that very minute; he had no return of the convulsions, his breathing became calm and easy, I ordered ten grains of Calomel every four hours, for three days, this produced ptyalism and the mercury was discontinued. At the time I write this he is so far recovered that he can converse rationally, can walk without assistance; his speech is not altered, except that he rather lisps from the loss of two front teeth – But Southover bells which have been ringing for the first hour the "Old Year out and the new year in" an immemorial custom in that village, have just ceased their chiming and warn me that it is time to retire. Old Ellis the Bellman, too, is crying past two o'clock. I must conclude.

He continued to visit Mr Moore frequently until he died at the end of the following year.

In the course of the eighteenth century the population of England and Wales had risen from five and a half million to over nine million in 1801 and to sixteen and a half million by 1831. This unprecedented increase was represented by a larger birth rate, a much reduced death rate, and the prolongation of the lifespan of adults. Even so in 1790 there was still a 25 per cent chance of infants not surviving beyond five years of age. Before examining the improvement in medical services other factors must be taken into account:

1. A change in drinking habits.
2. More abundant food.
3. Scientific advance.
4. Philanthropy.
5. Sanitation.

In the 1730s legislators had deliberately encouraged the drinking of cheap gin, immortalised by Hogarth with his depiction of the horrors of 'Gin Lane', as more corn was produced by the landowners and spirits were very lightly taxed. With time the appalling consequences gradually came to light and a series of hesitant steps were taken to increase the taxes on spirits and discourage their retail. Even after 1757 a fifth of the deaths in London were attributed to spirit drinking. After that, spirits gave way to beer and by the middle of the century tea had become a formidable rival to alcohol with all classes.

Advances in agriculture meant more abundant food for the many, though not for all. Scientific progress allowed superstition and the practice of presenting opinions in the absence of knowledge (sciolism) to give way to a more rational approach which was enhanced by a philanthropic caring for those previously neglected by society. Advances in sanitation had to await the nineteenth century.

Medicine had advanced with recognised courses for training, the provision of degrees and eventually the establishment of the Royal Colleges. Obstetric practices had improved with the training of accoucheurs. In 1757 midwife-based hospitals, for attending and delivering poor married women in their lying-in of their respective habitations, were established, with one to two man-midwives in reserve for problems. The midwives were paid 1/*6d* per birth, raised to 3*s* in 1767. As early as 1739 a special department for instruction in obstetrics had been created in the University of Glasgow. Pre-eminent in the teaching of midwifery were William Smellie (1763-97), Thomas Denman (1733-1815), William Hunter (1718-78) and Hunter's nephew Matthew Baillie (1761-1827). Denman studied medicine in London but obtained his MD from Aberdeen. He made £600 a year from his practice and £150 from teaching. David Daniel Davis (1777-1841) of Llandyfoelog, near Carmathen became the first professor of obstetrics. By 1800 the monopoly of midwives had been broken enabling obstetrics to be performed by general practitioners.

John Hunter did much to improve surgical practice. His brother William improved obstetric care, though personally resisting the use of forceps for assisting parturition. He kept a rusty pair which he claimed he never understood and never used. Their general approach, which Mantell seems to have followed, was very conservative, following Denman's law which states 'the head of the child should be rested for six hours as low as the perineum before forceps are applied, even though pains should have ceased by this time'. Denman preferred the use of the vectis (an obsolete obstetric instrument to release or 'carry' the head) to the use of forceps. Labour pains could be helped by opiate extracts, alcohol or herbal remedies. Religious opinion was against assisted childbirth. Prolongation of labour could be reduced by 'podalic version' – disengaging the head, turning the foetus and pulling on the feet; or by induction with oil, baths and enemas. Obstetric forceps were invented by the Chamberlen brothers in the sixteenth century but their secrecy was such that forceps only became generally available at the end of the eighteenth century.

Accelerated delivery occurred in only 3-5 per cent of childbirth. In 1737 the first caesarean section on a living woman in Britain was performed; though by 1800 just nineteen caesarean sections had been recorded in the British Isles with only two mothers and seven babies surviving. Induction by rupturing the membranes was called the English method. A finger was inserted into the dilated cervix sweeping the lower segment membranes off their loose decidual attachments to the uterine wall. High rupture of the membranes well above the internal cervical Os with a long metal catheter was employed by Hamilton in 1810.

Mantell used and wrote of the value of 'ergot of rye' to induce labour when this became available in 1827. Laennec introduced his stethoscope in 1827.

While forceps were being used by an increasing number in obstructed labour, extraction by craniotomy was still in use. Firstly the head was perforated by a craniotrite, and in some cases, the dead foetus was then left for a day to allow the body to shrink and so make for easier delivery. After the skull was opened crochets (hooks) or bony pliers could be applied. From this developed the cranioclast: one blade was passed inside the head and the other blade over the face. The use of these techniques to break the skull of the foetus if the head could not pass through the pelvis needs to be mentioned because such techniques reduced maternal deaths and allowed further pregnancies to occur. But perhaps the involvement, where necessary, of surgeons as well as midwives in delivery did most to improve standards of care.

The vectis, which Mantell possibly used, was like one curved blade of a forceps somewhat lengthened and enlarged. Later a tractor-vectis was introduced. Except in the most skilled hands it was highly dangerous with risks to the foetus and to the bladder. It was used to correct mal-position or aid the rotation of the head, and more dangerously as a lever or a traction instrument. Once well-warmed and greased, two fingers should be placed in the vagina to guide its application. According to the manuals of obstetrics at that time it could be used safely: to relieve face presentation with impaction, brow presentation to bring down the vertex, side of head presentation, cases of turning when the head cannot easily be extracted, if the labour pains were insufficient, or if convulsions occurred.

Mantell contributed to developments in medical literature most notably on fractures of the pelvis, the use of mercurial ointment in erysipelas, and on injuries to the head leading to hallucinations and suicide. He advised on the treatment of cholera and gave an early description of the treatment of greenstick fractures of the radius in children under nine years of age. Other short reports involved an enlarged thymus gland in asthma, an abscess of the prostate, a perineal fistula, and the treatment of enteritis.

He was one of the first to use the new drug ergot of rye in cases of prolonged labour, publishing his successful results in the *London Medical Gazette*. Ergot of rye is an oblong, slightly curved grain, about as thick and twice as long as a grain of wheat, of a dark brown external colour but lighter with a shade of pink internally. It was ground and given as a powder or more usually mixed with a little water or milk, allowed to simmer for a few minutes and given as an infusion, 20-30 drops of liquid extract every twenty minutes but repeated no more than four times. Used mainly to increase the strength of feeble labour pains, it was also recognised as able to originate uterine action, i.e. to induce labour. He wrote in *Lancet* on the treatment of erysipelas with mercurial ointment and described the occurrence of greenstick fractures in children under the age of nine.

Mantell's success as a surgeon – general practitioner – reached its zenith within a few years of taking over from James Moore. He received a salary from various parishes and from the barracks at Ringmer. For most other work there was a set price of payment when the patient could afford to pay. By superhuman efforts, many nights sleep curtailed, and devotion to his practice, his earnings became such that he could drive a hard bargain with any potential partner. There were other practitioners in Lewes, notably Thomas Hodson, who was both a colleague and a rival, but an older man who gradually reduced his commitment with age. After taking on first one then a second partner, Mantell concentrated on his more gentrified patients, leaving his partners concerned with the humdrum work. The local nobility looked upon him as their personal physician and more and more as a colleague. They paid him in friendship, in game from their estates, in dinner parties and theatre going, in aiding his publications and in securing fossils. With Mantell as their doctor they had little need to travel to London for a second opinion, Mantell could attract senior physicians and surgeons to Sussex, as anxious as anyone to see his fossil collections, and fully prepared to flatter him with their opinion that he was treating the prize patient in the best possible way. There is in fact reliable evidence that he was an excellent doctor and conferred with the best possible opinions keeping up to date.

His selective practice meant that he spent more time visiting patients and friends at Glyndebourne and Brighton on what was virtually a daily basis, and with partners taking the main workload he was able to attend meetings in London, Cambridge, Yorkshire and elsewhere, of the Geological Society, Linnaean Society, British Association for the Advancement of Science and the Royal Society. He spent more time visiting his relations, his wife's family and his publishers in London, travelling with his wife or with one of his children. He also increasingly spent weeks staying at various estates in other parts of the country, enjoying the company of others and using these resorts as a base from which to explore strata and quarries well away from Sussex. With his museum attracting many visitors, he spent the evenings improving his collection, talking to fellow geologists well into the night, and sometimes providing overnight accommodation and hospitality.

Whereas Mantell began his career as a successful medical practitioner but increasingly achieved fame for palaeontology with his discovery of the Iguanodon and other dinosaurs, James Parkinson, from whom he obtained considerable stimulation, was initially recognised for his three-volume compilation on fossils entitled *Organic Remains of a Former World*, and his survey of the ante-diluvian world. Most of Parkinson's fossils were acquired from dealers but he had assiduously gathered details of Georges Cuvier's studies in Paris and particularly his famous discoveries of giant extinct mammoths. His compendium was recognised as the first attempt in English to give a familiar and scientific account of the fossil relics of animals and plants, accompanied by figures of the specimens described. He was later to achieve universal distinction from his description of six patients with paralysis agitans, a degenerative disease of the nervous system, to which Jean Marie Charcot was later to attach the epithet, La Maladie de Parkinson.[11]

James Parkinson (1755-1824) had a very similar dissenting background to that of Gideon Mantell, which meant he could not qualify from a university. He was apprenticed to his father John Parkinson of Hoxton Square, Shoreditch and trained at the London Hospital, attending lectures by John Hunter before being awarded the diploma of the Company of Surgeons of London in 1782. The Royal College of Surgeons was not established until 1800. In the 1790s he was involved in politics as a member of the Society for Constitutional Information and also of the London Corresponding Society. The London Corresponding Society formed in 1792 was the leading radical political society in London with a total membership in 1795 of 3,000. Its aims were to forge links with provincial societies and to achieve a fair, equal and impartial representation of the people in Parliament.

Parkinson's involvement led him to produce a number of pamphlets under the pseudonym of Old Hubert. These included an address to Edmund Burke 'from the Swinish Multitude'. Following the arrest in 1794 of Thomas Hardy and others on a charge of high treason, Parkinson produced *Revolutions without Bloodshed or Reformation Preferable to Revolt* (1794) the proceeds from which were to go to the defendants' families. In September 1794 an attempt was made on the life of George III dubbed The Popgun Plot. The king had left Newmarket earlier than expected and so escaped the danger. Parkinson was summoned to appear before the Privy Council as he knew one of the defendants. Questioned about his membership of the committee Parkinson replied that he had been invited to join 'because they believe me firm in the cause (Mr Pitt smiled) of Parliamentary Reform, and because I had just produced a little tract for the benefit of the wives and children of the persons imprisoned on charges of high treason'. His later publications were mainly on medical topics such as observations on the nature and cure of the gout, and he became a trustee for the poor of the liberty of Hoxton and medical attendant to the private mad house at Holly House, Hoxton.

Mantell was not a person to drop past enthusiasms. On his return to Lewes to enter into partnership with John Moore as a surgeon, he made contact once again with his old mentor, the Rev. James Douglas (Anthony Brook, Memento Mori-3, October 2002) re-igniting his passion for antiquities and archaeology. Douglas, now in his sixties, had done most of his exploration thirty years earlier, although continued to encourage 'the methodical excavations of ancient monuments in the Lewes hinterland'. Even so these excavations were necessarily more basic than those of the Curwens and with each decade the scientific requirements of such explorations have become more and more demanding. His publications on antiquities in the *Gentleman's Magazine* and with illustrations in Horsfield's *History and Antiquities of the Town of Lewes* preceded those on fossils.

Interest in antiquities, the discovery of the artefacts of the past, earlier than the written word, with hard evidence from coins, ancient buildings, burial objects, etc. was a feature of the renaissance of the previous century. It arose in part from classical antiquity and as a reaction to the political and religious conflicts of the seventeenth century in which history could be written, rewritten and distorted. The London Dilettante Society set the pattern, promoting British understanding of Italian and later Grecian art and antiquities. Artefacts, such as coins offered a stronger foundation of truth. The study of the past – archaeology, history, antiquities – offered a means for those with leisure time to establish their identity in society combining a sense of intellectual endeavour with attention to the public good. Artefacts of historic or cultural value – tessellated pavements, stone circles, ruined abbeys – were recognised as public rather than private property to be preserved for others and later generations. Even so, the Antiquarian's acquisitive nature, wishing to form a personal collection, was often at odds with, and often overriding, preservation in the public interest.

The burgeoning of cultural life among the wealthy upper classes, seeping down to others with the spread of literature and cheap local newspapers, began in the eighteenth century in the Age of Enlightenment or Reason, as a European literary and philosophical movement ending with the French Revolution. It was rooted in the seventeenth-century scientific revolution and the ideas of Kant, Locke and Newton; marked by a passionate

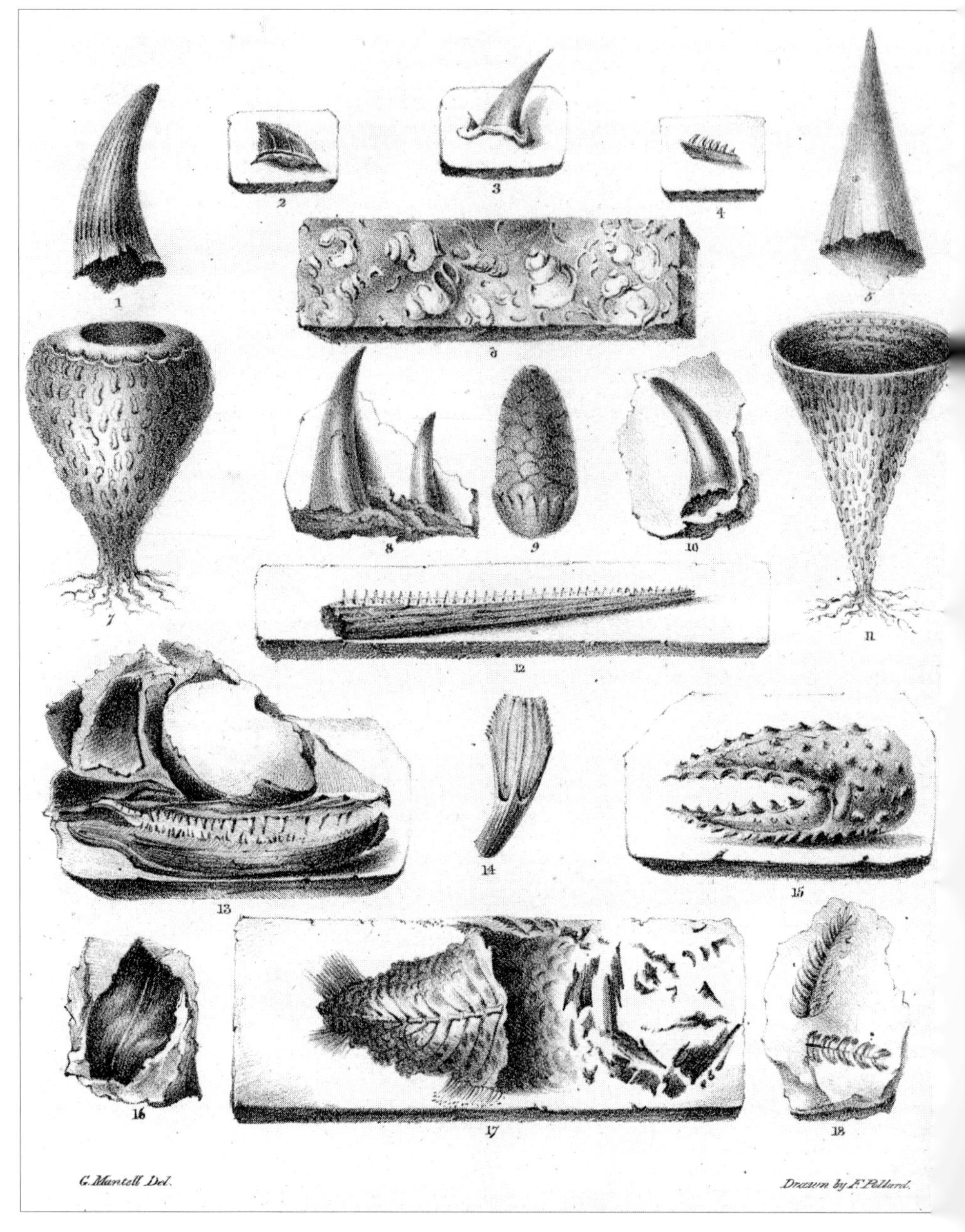

Fossils, as illustrated by Mantell, in Horsfield's *History of Lewes*.

concern with a theory of knowledge of the nature of things based upon a secure philosophical foundation. In Britain it continued in the early years of the nineteenth century, diffusing outwards from the major cities to the provinces with the increasing availability of information. People sought to establish their identity and that of the nation from examination of the past. The search for information led to a belief in progress, the formation of libraries, lectures and displays, and the gathering and collection of objects of the past. Nineteenth-century England was seen as the most bourgeois of nations as the middle class gained political power from the economic power of industrialisation, and to a lesser extent through the Reform Bill. For many, interest in antiquities and science offered more than economic reward, adding social mobility and the possibility of transcending class barriers. Although there was a downward spread of enlightened attitudes, the majority of the population were unaffected; economically for Britain its success largely depended on trade with the colonies involving sugar, tobacco and slaves.

Horsfield's *History and Antiquities of the Town of Lewes* (1824) and the interest it aroused contributed but one example of the growth of the allure for knowledge spreading throughout the kingdom. It contained an appendix by Gideon Mantell outlining the natural history of the environs of Lewes. In some twenty-two pages he outlines the geology and mineralogy with drawings of the shells and fossils found and then in sections on the zoology, ornithology, amphibian, ichthyology and botany listing the animals and plants with their Linnaean and English names and the local habitats where they are found. The massive tome was published by John Baxter and inscribed to William Baldock after whom Mantell's elder son was given his second name. Provincial societies were, if anything, more dynamic than those within the metropolis. Such studies were spearheaded by the lesser gentry, though they did receive a degree of royal patronage. The professions also became involved. Their elucidation appealed especially to lawyers. The clergy within universities were well equipped to collect artefacts and add them to newly formed museums. Within a small community, a clergyman was usually best placed to examine any antiquities found. Such study might advance an aspirant cleric, attracting preferment. The medical profession, surgeons and apothecaries, were more commonly associated with natural history but for those such as Gideon Mantell the wider social acquaintance with families of wealth, which their work inevitably entailed, led to similar opportunities to collect and itemise the data that artefacts provide.

Besides his interest from an early age in the discovery of fossils, Mantell collected rare plants and ferns. He gained recognition as a botanist, describing the fossil plant Alcyonium found in the chalk strata in a paper xx **APPENDIX.**

Linnean Names.	*Trivial Names.*	*Habitats, Observations, &c.*
Hirundo *urbica*	Martin	Builds in the eaves of houses in the High Street, County Hall, &c.[1]
* ———— *apus*	Swift	Builds in church steeples, castle turrets, &c.
* Caprimulgus *europæus*	Goat-sucker	On the authority of the late Mr. Woollgar.
Columba *turtur*	Turtle dove	In woods during the summer?
———— *palumbus*	Wood pigeon	In woods near Chailey, Newick, &c. rarely.
* Perdix *coturnix*	Quail	In the autumn on the Downs, where they breed; they are also found in broods in cornfields.
——— *cinerea*	Partridge	Common.
Phasianus *colchicus*	Phaesant	Common.
Otis *tarda*	Bustard	Sometimes on the South Downs, (W.)

WATER BIRDS.

Ardea *major*	Common heron	Occasionally seen in the Levels, prov. *Jack Hern.*
—— *stellaris*	Bittern	Several have been shot in the Levels.
—— *alba*	White heron	Very rare ;—One was shot by Mr. C. Bull.
Nunenius *arquata*	Common curlew	Frequent on the sea coast. In severe winters in the river and marshes.
———— *phœopus*	Whimbrel	Sea coast.
** Scolopax *rusticola*	Woodcock	Woods and rivulets.
———— *gallinago*	Snipe	Marshy and wet grounds.
———— *calidris*	Redshank	Sea shores.
Tringa *vanellus*	Lapwing	Downs and environs ; in moors and marshes during the breeding season, prov. *Pewit.*
*** Tringa *squatarola*	Grey sand-piper, or plover	In the winter on the coast.—rare.
** ———— *hypoleucos*	Common sandpiper	Banks of the river.
———— *canutus*	Knot	On the coast in August ;—shot by Mr. C. Bull.
Charadrius *pluvialis*	Golden plover	On moors and heaths ;—shot by Mr. C. Bull.
———— *morinellus*	Dotterell	Rare ;—shot by Mr. C. Bull.
* ———— *œdionemus*	Stone curlew	On the Downs[2].
Hematopus *ostralegus*	Common oyster catcher	On the sea shore, *prov.* "the Olive."
Rallus *aquaticus*	Water rail	In furze, cornfields, &c.
* Gallinula *crex*	Land rail	In furze, cornfields, &c.
———— *chloropus*	Water hen	Large ponds and marshes, *prov.* Moor-hens.
Fulica *atra*	Common coot	In ponds and marshes.
Alca *artica*	Puffin	On the sea shore, in the spring.
—— *torda*	Razor bill	On the coast.
*** Colymbus *immer*	Lesser imber	Mouth of the Ouse, December, 1823.
———— *auritus*	Eared grebe, or dubduck	In ponds and marshes.
* Sterna *hirundo*	Greater fern, or sea swallow	On the coast.
Larus *canus*	Common gull	On the sea shore.
—— *marinus*	Black-beaked gull	On the coast.
** — *hybernus*	Winter gull	On the coast.
** Procellarea *pelagica*	Stormy petrel	On the sea, very rare, one was shot by Mr. Bull.
*** Mergus *castor*	Dun diver	In ponds and marshes.—rare.
*** ——— *albellus*	Smew	Rare. A male was shot in the present year, (1823.)
*** Anas *cygnus*	Wild swan	Occasionally in severe weather ;—one shot by Mr. C. Bull.
—— *olor*	Tame swan	In the River.
—— *tadorna*	Shieldrake	Very rare ;—one was shot by Mr. C. Paine.
—— *anser*	Wild goose	In the winter, in the marshes and fens.

[1] In removing the eaves of a house in this neighbourhood, several old nests of martins were pulled down. In one of them the skeleton of a bird was discovered, still sitting, as if in the act of incubation ; and on removing it, several eggs were found beneath it. In all probability, the male had been shot, and the poor female, regardless of the calls of hunger, had continued at her post, till she expired!

[2] "The Stone Curlews live with us (in Sussex) all the spring and summer, and at the beginning of autumn prepare to take leave by getting together in flocks. They feed on earth-worms, that are common on sheep-walks, and downs : they breed on fallows." White's *Natural History of Selbourn.*

(Horsfield appendix) Mantell contributed an appendix to the Rev. Thomas Horsfield's *History of Lewes* (1824) providing a list of the flora and fauna of the surrounding area with their scientific and common names and the locality where they are found.

of June 1814, and was elected a member of the Linnaean Society before achieving recognition as a geologist. Travelling about his practice he noticed the flora including bee orchids, Nymphea alba; Nautilus, Anachusa, Toad flax, Saponaria (soapwort), Osmunda, charlock, etc., collecting some of these for his hortus sicca (xerophytic garden).

Stimulated by the Rev. James Douglas and others, he held a fascination for all manner of artefacts. In 1813 while digging out an arbour in the castle mound in his garden behind the surgery, he discovered a large conical earlier excavation containing an urn with the bones of a headless cockerel which he believed was a Roman 'augural relic'. Nearby were the skeleton of a boar, fragmented bones of a horse and a stratum of mussel and oyster shells amid fragments of charcoal. He opened a lead coffin in St Michael's church beneath the mutilated brass of a knight of the noble house of De Warrene. He paid workmen to dig into eight tumuli on Mt Caburn finding a green porcelain amulet and necklaces of jet and amber beads. Each skeleton had a small knife or dagger in the left hand. On Mount Harry, skeletons of those killed in the Battle of Lewes were discovered by workmen digging for flints. Brass Anglo-Saxon and Roman coins were found in nearby villages. At Lancing the pavement and other remains of a Roman Temple was discovered. And he purchased four brass armillae, a celt (a brass musical instrument) and a circular ornament which had been unearthed in the course of excavations at Hollingbury near Brighton. Road works to reduce the steepness of Malling Hill into Lewes from the east, led to the finding of skeletons of a Saxon burial ground. One of the skeletons had a spearhead and dagger. Mantell examined their jaws, commenting that these Saxons churls must have lived on a diet of corn or hard pulses to account for the worn-down appearance of the teeth.[12]

FOSSILS

> *He that enlarges his curiosity after the works of nature demonstrably multiplies the inlets of happiness* (Dr Johnson, quoted by Mantell).

A fossil is the imprint or remains of living forms; the word derived from the Latin, *fodere*, to dig and *fossa*, a ditch. The ancients, for example Herodotus, were aware of the entombment of shells in rocks and Leonardo da Vinci in the fifteenth century maintained that they were true shells and that the rocks containing them were once part of the sea bottom. The young Gideon said that he had read about them in the *Gentleman's Magazine*, probably based on an account of Brander's work on Hampshire fossils, *Fossilia Hartonensia*, originally published in 1766. More importantly, James Parkinson produced his three volumes of *Organic Remains of a Former World* in 1804, 1807 and 1811. That Parkinson was a definite inspiration to Mantell can be seen from Mantell's own statement:

> Having at an early period imbibed a predilection for the study of natural history and subsequently been educated in a profession fortuitously connected with that science, upon fixing my residence in Lewes I resolved to devote my leisure moments to the investigation of the Organic Remains of a former World, a study replete in interest and instruction.

Gideon wrote his first description of fossils, 'On the extraneous fossils found in the neighbourhood of Lewes' which appeared in the *Sussex Weekly Advertiser* in March and April 1812. That same year he was responsible for a private publication, *A Systematical Arrangement of Secondary Fossils* by G. A. Mantrell MRCS. The work contained a summary of Parkinson's work but also included a total of 38 fossil plants, 75 zoophytes, 44 echinoderms, 203 shells, 46 fishes and 7 arthropods which Gideon had personally collected. With these local publications behind him he felt sufficiently emboldened to travel to London with the express purpose of meeting James Parkinson and hoping to acquire a copy of his works.

It was not only the fossilised bones which attracted attention. John Shute Duncan wrote that 'The Noble Science of Geology / Is fully bottomed on Coprology'. In the same strata as fossils, coprolites were dispersed irregularly and abundantly, representing the petrified faeces of extinct creatures containing remnants of their meals such as the scales and occasionally the teeth and bones of fishes.

The finding, buying, exchange and collection of fossils in the nineteenth century was the vogue of the time, just as stamp collection became in the twentieth century. There were many collectors in Lewes, including James Moore, John Tilney and Thomas Woollgar. Mantell exchanged fossils from Cornwall, Bristol, Wiltshire, Norfolk and elsewhere; and in London, with well known dealer John Stutchbury. Mantell was anxious to exchange fossils, receiving finds from the continent and America. Robert Bakewell (1768-1843)[1], author of *A Popular Introduction to Geology*, was at sixty years of age regarded as an elderly curmudgeon. Yet Mantell perceived him as a most intelligent and agreeable man, and invited to preview his collection, dubbed the Mantellian Museum, before its opening. Bakewell was duly impressed and wrote a letter introducing him to a key correspondent in America, Professor Benjamin Silliman, suggesting the exchange of Sussex fossils with those from America. Mantell belatedly wrote to Silliman six months later in April 1830, sending him copies of *Illustrations of the Geology of Sussex* and his museum catalogue. This letter crossed with one from Silliman and led to a notable succession of correspondence up to Mantell's death. At the time Silliman was fifty-two and Mantell forty-one.

Surprisingly they did not meet in person until 1850. In these letters Mantell expresses his worries, to which Silliman provides an optimistic reply. Yale University Library contains 126 letters from Mantell and 94 from Silliman (returned to him after Mantell's death). The letters covered a range of subjects. Early on there was an exchange of family details. Mantell's personal health, his ambitions and predicaments were frequently mentioned and both ventured into political debate. In June 1833 Mantell wrote that 'we are here in great consternation lest the Duke (of Wellington) should again get into power; if he does your country will be the only asylum'. Silliman replied the following month: 'it is probable that slavery will in future be the principal source of discussion'. In 1849 they discussed the American War of Independence:

> I fully reciprocate your feelings as to the virtues of our blessed monarchs of the 17th C. You forget that my family was despoiled of lands and fortune and station for their attempts in the previous century to maintain the Protestant faith ... and in Charles II's time when the persecuted family took refuge in Lewes, their names are among the fined and imprisoned for being conventiclers.

Professor Silliman (1779-1864) was born in Connecticut. His paternal family originated from Italy, but after the Reformation a Sillimandi left Tuscany for Geneva and one of his ancestors went over to America. On his mother's side were pilgrims of the *Mayflower*. His great great grandmother, Priscilla Mullins, married John Alden as commemorated by Longfellow in his poem 'The Courtship of Miles Standish'. Silliman was originally admitted to the bar but in 1802 became lecturer of Chemistry and Natural History at Yale College (it became a university in 1887). In 1805-6 he travelled to England, Scotland and Holland. On a trip to Europe in 1817 he developed a particular interest in mineralogy and geology, and started the first geological course at any American institution in 1833. In 1818 he founded the *American Journal of Science and Arts*, and remained its editor in chief until 1856. This correspondence, along with the diaries, provides a remarkable insight into Mantell's thoughts over many years.

We owe to William Smith (1769-1839), a surveyor and civil engineer involved in coal mining and later building canals, the first appreciation of the stratification of rocks. Though self-educated he grasped and applied some fundamental principles to modern geology, namely that rock materials deposited as sediments form strata, and that these strata become fashioned into recognisable and predictable sequences over wide areas. The individual strata can be identified and correlated on the basis of the unique fossils found within them, and from the fossil contents rocks of a similar age could be identified. He made the first attempt at a geological map in 1796 and circulated an account of his method for tracing the strata by examining the organised fossils embedded within them. Starting with a geological map of the district near Bath, he produced in 1801 a *General Map of the Strata of England and Wales*. In 1808 he was commissioned to supervise the extension of the Ouse navigation in Sussex and over the next four years he extended it a further 11.2 km (seven miles) inland from Sheffield Bridge to the Balcombe road near Cuckfield. He also submitted a plan to the Clerk of the Peace at Lewes for a branch to join the Grand Southern canal near Crawley. In 1809 while searching for materials to construct locks and bridges he went to a quarry near Cuckfield and was shown very large bones later identified as those of the Iguanodon, a decade before Mantell. In 1815 he enlarged his findings, mapping the

Delineation of the Strata of England and Wales, exhibiting the collieries and mines, the marshes and fenlands originally overpowered by the sea, and the varieties of soil according to the substrata. With twenty different tints used to outline the boundaries of the strata he bankrupted himself in paying for the production and went to live with his nephew, John Phillips, in Yorkshire. He received a Doctorate of Laws from Dublin University in 1835 and chose the stone for the Palace of Westminster working with Charles Barry and H. T. De la Beche. Professor Buckland, the principal exponent of the new science of 'undergroundology' published the first comparative table of the strata of England in juxtaposition with that of the Continent.

Scientists had to battle against the religious dogma of the day; according to Bishop Ussher of Armagh in 1850, God made the world 4,004 years before the birth of Christ and that in the story of the Creation all creatures were made the same day. To explain fossils and extinct species intellectuals and scientists cited Noah's flood whereby the deluge destroyed all creatures except those saved by Noah in his Ark.

If William Smith was the founder of English geology, the Rev. William Buckland (1784-1856) was its first highly eccentric showman for 'undergroundology' as he termed it. He was later to greet Darwin's *Origin of Species* with the remark: 'I am not quarrelling or finding fault with a crocodile, a crocodile is a very respectable person in his way, but I quarrel with finding a man, a crocodile improved.' With his almost blind father he had explored the quarries and woods of Devon and found fossil sponges in the chalk downland. He reassured his audiences that the facts of geology are consistent with the record of the Bible and equated with Noah's flood; as a clergyman he was dedicated to confirm their compatibility. To Buckland, geology was a master science through which he could investigate the signature of God. His treatise of 1823 was entitled *Reliquiae Diluvianae* (or Relics of the Deluge). He was reader in mineralogy and later of geology at Oxford before becoming Dean of Westminster in 1845. Of medium height, good looking and given to eccentric dress, he could be observed wearing a black academic gown when in the field as though he was a necromancer. His carriage, like the corridor to his Oxford rooms, was laden with rocks, shells and bones in dire confusion. Visitors to his rooms in Oxford would also meet dressed up in cap and gown a pet bear named Tigleth Pileser after the powerful king of Assyria to whom Ahaz turned for help against Syria and Israel. Other creatures in his rooms according to an Oxford tutor, William Stanhope, were snakes, green frogs, guinea pigs and a jackal. If that was not all, to add to his eccentricity he planned to eat his way through the animal kingdom.

His lectures attracted a vast audience and were full of bustling eloquence, cheery, ebullient humour, wit and anecdotes. He would discuss the bones of exotic animals found in Britain including hyenas, marine reptiles and opossums discovered from among Jurassic rocks. Like any true scientist he was prepared to test his hypotheses. In 1821 quarrymen found a cave at Kirkdale in the Vale of Pickering, Yorkshire, full of fossilised bones. Believing that these bones had come from a den used by hyenas, as in tropical climes with a wide range of species including lions and tigers, he imported 'Billy' an African hyena and compared the gnawed marks on the bones which it ate with similar markings on bones from the cave. Furthermore he gathered countless hyena canine teeth from the cave and concluded that the ancient hyenas were a third larger than the modern specimen.

He examined and described the 'Red Lady of Paviland' found in a cave in South Wales, believing her (actually a male) to be a camp follower of the Romans. However, once stratification of rocks was recognised with different types of fossils found in particular layers, the idea of a single deluge no longer 'held water'. However, Buckland's keenness to reconcile science with religion won him support in high places: from Robert Peel, Sir Everard Home, Sir Joseph Banks the botanist, and Lord Grenville, Chancellor of Oxford University. Not only did he lobby for the first chair of Geology at Oxford but secured a stipend from the Treasury and became Director of the Ashmolean Museum. Buckland's dilemma was immortalised within the university in a popular satire, *Facetiae Diluvianae*, in which he met Noah and each added to the bewilderment of the other.

James Hutton (1726-97) an Edinburgh graduate had studied law and medicine, becoming an MD from Leyden University and involved in the manufacture of ammonium chloride, farming and navigation between the Forth and Clyde before concentrating on geology. Later known as the gifted interpreter of the Divine Power and Wisdom at work in the Laws of Organic Life, in his three-volume *Theory of the Earth, with Proofs and Illustrations* (1795-97), originally read to the Royal Society in 1785, he described a series of floods with an endless cycle of forming and reforming the landscape, cycles in which the rivers carried sediment from the land to the sea. The earth, its mountains, plains and waters were endlessly remodelled. Rocks were igneous, metamorphic and sedimentary. Mountains arose through orogenesis, were weathered and eroded; the sediments washed down by rivers became alluvia. The soil could be covered by wind-blown loess. The landscape was subject to synclines and anticlines and the coasts were reshaped by erosion, shingle, longshore drift and inundation. He saw no need to invoke catastrophic causes in understanding the earth. Its former history could be adduced from a study of the processes observable in the present. He noticed strata of sedimentary rocks lying at an angle to each other, originally laid down horizontally on the ocean floor, then elevated, folded, the top of the folds eroded, subsequently sinking back into the ocean. Such changes occurred over millions of years. His language was somewhat diffuse and failed to gain recognition until the mathematician, John Playfair, in his *Illustrations of the Huttonian Theory of the Earth* (1802) produced a more lucid and fresh exposition of the theory, quoting Isaac Watts ('O God, Our Help in Ages Past'): 'Time like an ever rolling stream.'

A verse of Tennnyson's *In Memoriam* reflected Hutton's view:

> The moanings of the homeless sea,
> The sound of streams that swift and slow,
> Draw down Aeonian hills, and sow
> The dust of continents to be.

In Edinburgh Hutton would have come under the influence of Adam Smith's *Wealth of Nations,* suggesting the interplay and cyclic nature of economies. John Stuart Mill even attempted to draw parallels between geology and the law:

> The deposits of each successive period are not substituted but superimposed on those of the preceding and in the world of law, no less than in the physical world, every commotion and conflict of the elements has left its mark behind in some irregularity of the state; every struggle which ever vent the bosom of society is apparent in the disjoined condition of the past. In the field of law which covers the spot may the very traps and pitfalls which one contending party set for another are still standing.

Geology had become a leisured pursuit of the intelligentsia in the late eighteenth and early nineteenth centuries, however, as Samuel Butler records in *The Way of All Flesh*, geology meant little to the average youngster at the start of the Victoria's reign. Not until Louis Agassiz, in 1840, brought to Britain the idea of glacial valleys and moraines formed in the Pleistocene era when great masses of ice and subsequent glaciers covered much of the land, *only then* was there any acceptance of the gradual scheme of environmental change with its revisions and readjustments.

In the same vein, Parkinson, a founder member of the Geological Society, had written that the account of Moses in the Bible is confirmed in every respect, except as to the age of the world, and the timescale between the completion of different parts of creation. The formation of the globe and the creation of life must have been the work of a vast length of time. He maintained, as had the Swiss geologist, De Luc (1778), that each of the six days of the Creation represented a geological epoch; thus the word 'day' was used to designate indefinite periods in which a particular part of the great work of creation was accomplished. In defiance of the conventions of that time, he agreed with Hutton that nature was in a constant process of transformation and the earth would continue to undergo regular changes.

James Parkinson's *Organic Remains of a Former World* was a remarkably ambitious attempt to record and depict all known British fossils, as part of a larger endeavour 'to find out the ways of God in forming, destroying and reforming the Earth'. Their publication was described by Thackeray as the outstanding event in the history of our scientific knowledge of fossils. Volume I was of vegetative fossils, discussing the process of fossilisation and the great flood; volume II concentrated on fossil zoophytes; and volume III on star fish, shells, insects and mammalia. To enlarge upon the mammalian fossils, Parkinson had assiduously gathered details of Cuvier's famous discoveries of the giant extinct mammoths, with others found near the Hague in Holland and rescued after the French Revolution. Mantell made a special trip to London to meet Parkinson; he was nervous with excitement at the pleasure and privilege of the anticipated meeting. Parkinson, then fifty-eight years old, greeted Mantell courteously with a mildness of manner. He was rather below medium stature, a pleasant personality with an energetic intellect. This encounter transformed Mantell from a casual hobbyist into a serious and original investigator of fossil life. The twenty-one-year-old Mantell was impressed but unfortunately unable to afford to buy the volumes. However, as stated, Mr Woodhouse, his would-be father-in-law, bought a copy for him.

New scientific concepts had to battle against the theocracy, public opinion and even the leaders of the scientific establishment. Such concepts were castigated as distortions of science into a sophism against the Scriptures of eternal truth. It did not help that the original evidence came from France. Cuvier had found that the mastodon, Megatherium and mammoth were species distinct from present-day elephants.[2] The bones of these creatures had been found in Holland and brought to Paris. Later a virtually intact mammoth was found in the Siberian tundra. Through his examination of 'crocodile' fossils from around Honfleur and Le Havre he concluded that these came from a different layer of great antiquity considerably older than those containing the bones of mammalian quadrupeds. Successive forms of life found in the fossil collections resulted from serial episodes of creation and extinction. The doctrine of catastrophes, of a series of revolutions wiping away former worlds and destroying ancient forms of life, with each layer of rock holding its own characteristic fossils, had still to be universally accepted.

Georges Cuvier (1769-1832) republished some of his earlier work in four substantial volumes as *Recherches sur les Ossements Fossiles de Quadrupèdes* in 1812. Volume I discussed the Egyptian ibis and gave a geological description of the environs of Paris. Volume II detailed living and fossil rhinoceri, hippoptomi, tapirs and elephants. Volume III described two large fossil animals Paleotherium and Anoplotherium found in the vicinity of Paris with techniques to be applied: 'The first thing to do in studying a fossil animal whether it be a carnivore or a herbivore; if the latter, one can then ascertain to which order of herbivores it most nearly belongs'. He gave anatomical notes on the Irish elk, fossil deer, aurochs and other bovines, horses, pigs, bears, dogs, cats, sloths, dugongs and seals; recognising that the climate of Europe had changed greatly since then or that the animals habitat preference had changed. In volume IV he surveyed fossil bones from oviparous quadrupeds including crocodiles, turtles and lizards (saurians). All these fossil crocodiles occurred in strata significantly older than those containing the remains of Paleotherium and Anoplotherium.

Cuvier's writings disseminated through the geological world; thus Buckland travelling in Dorset by coach noticed a lady reading the same tome. 'You must be Miss Morland,' he tentatively presumed, 'to whom I am about to deliver a letter of introduction.' Although he was a bachelor of over forty years, he speedily married Miss Morland and the wedding tour was a geological excursion across Europe over the next year. In a letter of 17 April 1813, soon after the appearance of Kerr's translation of Cuvier's work, the Rev. James Douglas advised Mantell that the 'great desideratum of geology is now fixed on strata in which the bones of quadrupeds are found, they determine the last great revolution of our planet antecedent to the present order of created life.'

One person who was quick to accept the 'notion' of Cuvier that the world had been destroyed several times before the creation of man, was Lord Byron (1788-1824) in his poem 'Cataclysms':

> When this world shall be former, underground,
> Thrown topsy-turvy, twisted, crisp'd, and curl'd,
> Baked, fried, or burnt, turn'd inside-out, or drown'd,
> Like all the worlds before, which have been hurl'd,
> First out of, then back again to chaos,
> The superstratum which will overlay us.
>
> So Cuvier says, – and then shall come again
> Unto the new creation, rising out
> From our old crash, some mystic, ancient strain

> Of things destroy'd and left in airy doubt;
> Like to the notions we now entertain
> Of Titans, giants, fellows of about
> Some hundred feet in height, not to say miles,
> And mammoths, and your winged crocodiles.

To counter the doubts which continued to be made all too vocally by the establishment and others, Mantell found it necessary to introduce his major work on *The Fossils of the South Downs* in 1822 with a preliminary essay of thirteen pages presenting an earnest apologia written by the Reverend Henry Hoper of Portslade, himself a fossil collector, 'On the Correspondence Between the Mosaic Account of the Creation and the Geological Structure of the Earth'. The essay included appropriate texts from the Book of Genesis:

> 'Let the waters under the heavens be gathered together into one place and let dry land appear'
> 'The earth was without form and void'
> 'The Spirit of God moving upon the waters'
> 'The earth was corrupt before God and the earth was filled with violence'

He explained that the waters at one time covered the earth and the content of the earth lay beneath the waters, but not in their present form. The earth's crust was formed from water and there were a series of chaotic events previous to the present order of things. The immense accumulation of fossil animals, he believed, could not be attributed to anything less than the absolute destruction of whole orders of creation. He ended his introduction declaring that Moses was not entering into philosophical inquiries but to claim for the God of Israel the glory of having created the whole visible universe and to give an account of the origin of man.

There was a letter quoted by Spokes from an unknown gentleman of Norwich to Mantell in July 1831: 'We are led by the Sacred Scriptures to believe that there will be a resurrection of our bodies and that we shall appear in a material form. As this is to be the case why may not the human remains of the former world would have been raised in like manner?'

Mantell presented data of the South Downs and Weald in great detail with each locality described; a result of many field trips to verify his findings. The tome with its maps, illustrations and forty-two fine plates involved considerable application over a decade of study.

Dedicated to Davies Gilbert MP, he was very proud of the forty-two plates by Mrs Mantell which were her first attempts at lithography:

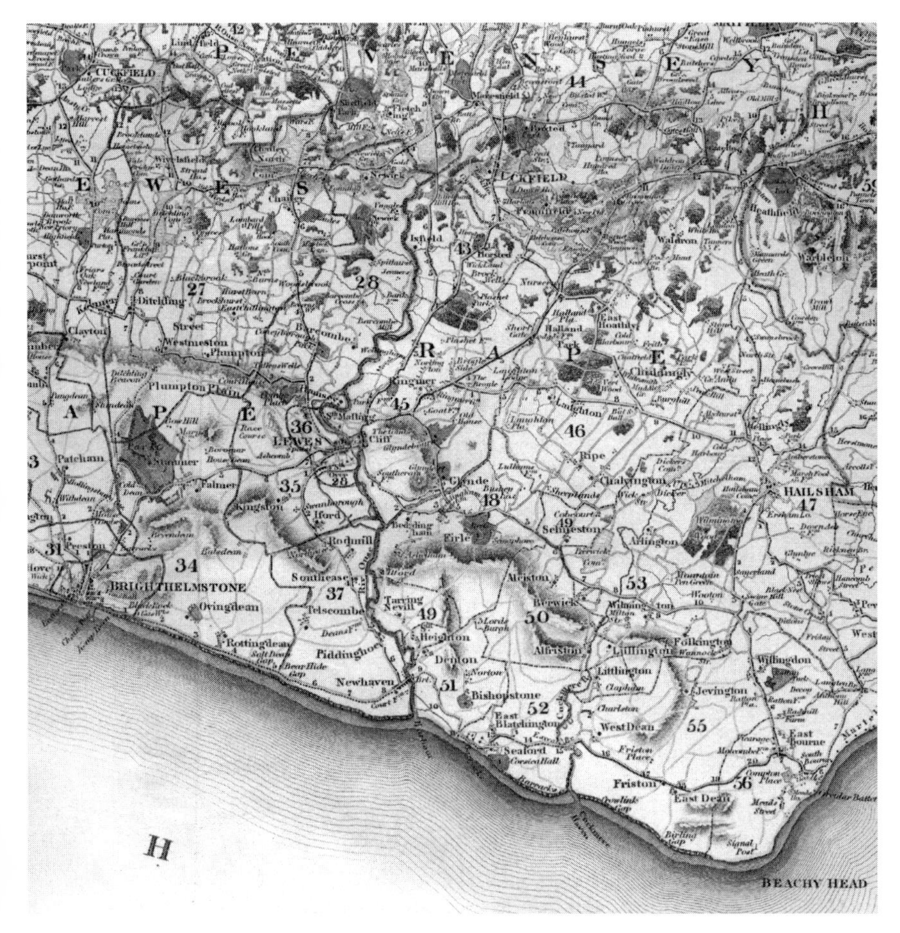

Excerpt of a map of Sussex in 1825 by Greenfield from a survey of 1823, showing Cuckfield, Brighton, Lewes, Ringmer and Laughton.

> As the engravings are the first performance of a lady but little skilled in the art, I am most anxious to claim for them every indulgence. I am aware that the partiality of a husband may render me insensible to their defects, but, although they may be destitute of that neatness and uniformity which distinguishes the work of the professional artist, they will not, I trust, be found deficient in the more essential requirements of correctness.

Mrs Mantell's contribution was reinforced by a poem from Gideon's friend, Horace Smith:

> My dear Mrs Mantell 'tis monstrously hard
> When my head's full of Monsters to think that the Bard

To your Album can make contribution
Your Museum has banished the Muse, and a train
Of Giants and Gorgons bewilders my brain
With a phatoasmorgoric confusion.

Mantell used the newly developed camera lucida to sketch the outlines of the Downs. To pay for its publication, subscriptions had to be sought in advance; this could be a most difficult procedure. All in all, it was an extravagantly handsome volume reflecting Mantell's social aspirations. As a work of science, *The Fossils of the South Downs* unquestionably succeeded, but it failed to achieve either the popular or the social acceptance for which the author had hoped. John Phillips, in his appreciation of the work, said that until the appearance of Dr Mantell's work on the geology of Sussex, the peculiar relations of the sandstones and clays of the interior of Kent, Sussex and Hampshire were entirely misunderstood.

When Nature was young, and the Earth in her prime,
All the rocks were invited with Neptune to dine
On his green bed of slate he was gracefully seated
And each as they entered was civilly greeted.

Oolites, with sandstones and sand red and green
In a crowd, near the top of the table were seen,
The last that were seated were Chalk marl and Chalk
They were placed close to Neptune to keep him in talk
Now the God gave his orders, 'If more guests should come
Let them dine with the lakes in a separate room'.

Mantell listed the various strata of the South Downs and Weald:

1. Ferruginous sand, with subordinate beds of sandstone, etc.,
2. Blue clay, Oak Tree soil, including Sussex Marble (formed from Paludrina),
3. Brick earth,
4. Greensand,
5. Chalk, Blue Marl, Chalk Marl, Lower or Hard Chalk, Upper chalk,
6. Plastic clay,
7. Blue clay of the levels,
8. Alluvium – mud of rivers, deltas, gravels of torrents, the effect of causes which still continue in activity; and Diluvium, gravel, boulders, sand, produced by cause no longer in activity.

In ferruginous sand he found the teeth of fishes and alligator like animals and the bones of the latter; in blue marl – crabs, fishes, corals, belemnites, ammonites, inocerami; in chalk marl – vertebrae and scales of fishes, shells, shark's teeth. He can also be credited with understanding the irregularities which occur in the strata, their inclination and dislocations clearly displaying to geologists that they had been subjected to great violence.

One result of the publication of his *Geology of Sussex* was that others sought to obtain specimens from the quarries paying the workmen more than their previous payments. He went with his brother Joshua to Portslade on a visit to the Rev. H. Hoper who had several splendid specimens of fossils which he had obtained from 'my workmen at Lewes chalk-pits; this kind of poaching has become so general since my work has appeared that I have now no hope of adding anything interesting to my collection.'

One unusual aspect of the strata was the thin seam of Sussex marble running from Bethersden in Kent as far as Petworth in West Sussex. Mantell 'discovered a new habitat of Sussex marble at Plashett's farm' in Ringmer parish some twenty feet below the surface in a bed of blue marl and clay. He also reports when visiting Col Downman at Laughton Lodge that large blocks of Sussex marble were dug up during the sinking of the foundations of the house. The seam is rarely more than a few feet across; it is formed from the fossils of Paludrina viviparous, a freshwater mollusc, in metamorphic limestone rock, possibly due to igneous intrusions. The fossils crystallise and are then destroyed. Though not a true marble it can be worked and polished, used for fonts and pillars in parish churches, street paving in Biddenden, and mantelpieces in farmhouses. Gideon Mantell describes how

> two pieces of slate (5ft by 4) of Sussex marble dug up at Laughton were sent to Parsons to be cut and polished. The works at Laughton were instituted by the Earl of Chichester from perusing the notes in my work on the Geology of Sussex; they give employment to upwards of 30 poor persons, who would otherwise have been destitute of work at this severe season, if my work be productive of no other practical good but this, still I shall have the satisfaction of feeling that my labours were not entirely useless.[3]

Visitors to Mantell's collection would appear unexpectedly. Thus on the evening of 4 October 1821 a young man appeared, of medium size with small eyes, fine chin and a broad expanse of forehead, and introduced himself as Charles Lyell of Bartley Lodge, Herts., a pupil of Professor Buckland of Oxford. He had been visiting his former school at Midhurst in West Sussex when quarrymen told him of a 'monstrous clever man, as lived in Lewes ... who got curiosities out of the chalk pits to make physic with'. Lyell was so intrigued with their account that he rode for twenty-five miles across the Downs to track the man down. After examining the collection he stayed to drink tea and discussed geology till past midnight, returning two days later. He informed Mantell that the late Dr Arnold was an intimate friend of his, and used to be continually referring to Mantell's paper on the Alcyconium. The visit to Mantel enthused and motivated him with the subject so much that he took it up in preference to law.

Charles Lyell (1797-1875, later Sir Charles), was perhaps the most intellectual of the geologists. The consequent outcome from the meeting was a correlation of Mantell's secondary formations in Sussex with those further West. The young Lyell had travelled widely both in England and on the continent and acted as interpreter when Cuvier came to England. Later he managed to arrange a return visit to France when he tried to convince Cuvier of the 'geological deluge' but Cuvier and Prevost convinced Lyell that the Parisian formations could have been deposited under conditions similar to those of the present seas and lakes. In the first volume of his *Principles of Geology*, July 1830, – 'the most philosophical geological work that had appeared' – he developed a rough timescale for the most recent era of the earth's history as represented by the tertiary formations and attempted to explain the former changes on the earth's surface by reference to causes now in operation with systematic descriptions of volcanoes, earthquakes, sedimentation and erosions.

When he was Professor of Geology at King's College, London, Lyell visited many countries in Europe to examine their strata and fossils and to persuade them of his views. His *Principles of Geology* has been described as a forceful criticism of other geologists thinly disguised as a history of geology, attacking those who arbitrarily invoked catastrophes at every turn and those claiming scriptural authority to limit the timescale of the earth's history.

> Amid all the revolutions of the globe the economy of Nature has been uniform and her laws are the only things that have resisted the general movement. The rivers and the rocks, the seas and the continents, have been changed in all their parts but the laws which direct those changes, and the rules to which they are subject, have remained invariably the same.

He was later to urge Darwin to publish on evolution but stopped short of public support for him. But more importantly, with his *Antiquity of Man* (1863) he established that human beings lived alongside the mammoths and other extinct mammals of the glacial period – known by Lyell's designation the Pleistocene – and finally accepted evolution by natural selection. Mantell came to rely on Lyell for support in later controversies.

Mantell was initially interested in the fossils of the chalk Downs: bivalves, sponges, and molluscs such as belemnites. His description of the fossil Alcyonium, a funnel-shaped sponge once thought to be a coral came to the notice of George Bellas Greenough, founding President of the Geological Society. He also sought the help of James Sowerby to identify the invertebrates he had found. Sowerby was a naturalist determined to catalogue British fossil shells in a work produced in serial form: *The Mineral Conchology of Great Britain*, started in 1812. After exchanging letters in 1813, Mantell sent Sowerby specimens of Scaphites, Turrilites and Ammonites from the marl pit at Hamsey, a hamlet to the north-east of Lewes. In return for the perfect specimens Mantell sent, Sowerby named one species after him, *Ammonites Mantelli*. Mantell's contacts with Greenough, Sowerby and Aylmer Bourke Lambert, a botanist, led to his election to both the Geological and Linnaean Societies. To his great satisfaction he succeeded in his election to the Linnaean Society but through a clerical error and the illness of its Secretary, Thomas Webster, did not hear of his election in May 1818 to the Geological Society for two years. The encounters with these naturalists gave Mantell a better understanding of fossil plants as well of the species of invertebrates.

The Geological Society of London had been formed as a dining club in 1807 but grew into a scientific society to 'make geologists acquainted with each other, stimulate their zeal and contribute to the advancement of Geological Science', with premises at No. 20 Bedford Square. Although the Geological Society had a large number of members, Mantell commented that 'there are very few who cultivate geology as a science. They belong because it is not difficult to obtain admission and FGS is a fashionable title.'

As a society of the elite neither Mary Anning nor William Smith, the surveyor, were admitted. Robert Bakewell, an amateur geologist who wrote *A Popular Introduction to Geology* was also excluded, writing that 'there is a certain prejudice among members of the scientific societies which makes them unwilling to believe that persons residing in provincial towns or the country can do anything important for science'. Mantell also experienced difficulties before becoming a member of its council in February 1825. In a letter to William Fitton[4] he had described a freshwater mollusc, *Unio valdenisis*, a clam or pearl mussel, found in the Tilgate beds. Greenough an FRS, former MP and President of the Geological Society was convinced that the iron sand was always a marine deposit. Over six months elapsed before the paper was read to the Society, and it remained unpublished for three years. Buckland shared Greenough's misunderstanding and wrote to Mantell to warn against claiming that the teeth and bones were found in the older iron sand formation.

However, in 1822, the dispute was settled when Fitton visited the Tilgate beds with Gideon Mantell and accepted his premise. He agreed that what Mantell had called undulating furrows in the sandstone of Tilgate forest were ripple marks resembling the patterns found in sand on a beach. From this he speculated that the rocks might have been found in a low-lying flood plain or along the edge of a lake or river delta where sandbanks accumulate. He further recognised that Sussex marble was formed from the shells of a freshwater snail, *Paludrina viviparous*; and between them they added other freshwater species.

Acknowledgement of the discovery of the freshwater origins of the Wealden strata emerged as Mantell's most significant contribution to the understanding of the geology of Sussex. With the *Illustrations of the Geology of Sussex* in 1822 the full value of the evidence became clear, explaining the relationship of the sandstones and clays of the interior of Kent, Sussex and Hampshire. He contended that the calciferous sandstones of Rye, Winchelsea, Hastings, Tilgate forest, and Horsham were but different portions of the same series of deposits, belonging to the iron-sand deposits formation; the Ashburham beds being situated beneath them. He believed that the vegetation and animals of the Tilgate strata must have been overwhelmed 'by a fluid in a state of violent commotion', since they were generally broken and their fragments promiscuously intermingled. But the perfect manner in which teeth, scales and other parts were preserved there gave reason to suppose that the originals were not transported from a distance but the animals lived and died in the vicinity of the district where their remains were entombed. All of which indicated the existence of dry land at no great distance from the borders of a river or lake. In addition to his reptilian fossils, Mantell was responsible for the recognition of seventy-two new species though some were subsequently modified. There were four new genera of zoophytes and new species included fifteen zoophytes, four echinites, twelve ammonites, three hamites, two scaphites, twelve belemnite, one turrilite, nine univalves, seven inocerami, nine bivalves, one crustacean and four new fishes.

He received help from a variety of people throughout Sussex, thus Mr Drewitt of Peppering near Arundel wrote to him after finding the remains of a fossil elephant in his garden. He encouraged some of the younger geologists such as Frederick Dixon who died in 1849. His work *The Geology and Fossils of the Tertiary and Cretaceous Formation of Sussex*, edited by Owen, was published in 1859. He stated that: 'In Sussex

Dr Mantell has been most conspicuous in the advancement of geological knowledge and as the historian of the Weald formation and discoverer of the Iguanodon, his name is placed among the first geologists of every age and country'.

DISCOVERIES

Discoveries depend on circumstances; in the words of Szent Gyogyi 'of seeing what everybody has seen but thinking what nobody has thought' – to discern what lies behind the camouflage. The Maastricht animal, the Mosasaurus, was the first prehistoric reptile recorded. In 1786 when first discovered it was believed to be a whale, later considered to be a crocodile and, when examined by Cuvier in 1808, was classified as saurian.

At Lyme, Richard Anning, a local carpenter, combed the coast and displayed the fossils he had found in order to augment his income. Jane Austen may have bought various relics from him during her visits to Lyme in 1803 and 1804, the town where she wrote her novel *Persuasion*. Sadly, Anning fell down a cliff and later died of tuberculosis in 1810, leaving a widow and two young children. His wife, Mary Anning (1799-1847) as a young girl had been struck by lightning when held by her nurse. She was unhurt but the nurse died. Mary Anning eked a living, continuing to collect and sell marine fossils from the sea cliffs at Lyme Regis thereby lifting her family out of poverty. Buckland, Greenough, and Conybeare were among the eminent geologists who bought her fossils. She developed a genius for discovery and piecing together the fossils with the ability, judgment and skills to extract fossils imbedded in the Lias – a corruption of layers, referring to alternate beds of limestone interspersed with clay and marl. The local landlord, Lt-Col. Thomas Birch, (later to be called Thomas Bosvile), discovering the family's plight, arranged an auction on their behalf and obtained high prices for the fossils.

As she became more adept at knowing where to look, Mary Anning uncovered several species of Ichthyosaurus. These fish-lizards possessed the snout of a porpoise, head of a lizard, teeth of a crocodile, vertebrae of a fish and paddles of a whale. They were air-breathing, cold-blooded and carnivorous. Her greatest discovery followed the finding of a large head by her son, Joseph. She eventually unearthed the entire seventeen-foot skeleton of a marine reptile which came to be known as the Plesiosaurus (1821). It had a serpent-like neck nine feet long, as long as or even longer than the body and tail combined with 30-49 vertebrae. Conybeare was ecstatic at the news and later received from Mary Anning a drawing of the specimen. When a similar drawing was shown to Cuvier he dismissed it as a fake, on account of his a low opinion of English anatomists such as Sir Everard Home, in view of the ridicule caused by Home's five contradictory papers, rushed into print, on the 'Proteosaurus'. A special meeting was convened to examine the specimen and its genuine nature confirmed by Conybeare and Konig. The identities of both Ichthyosaurus and Plesiosaurus were resolved in a paper by Conybeare and De la Beche in 1821, recognising them as oviparous marine quadruped lizards, Enalio-sauri.

In 1823 She also discovered a flying reptile, the pterodactyl, described by Buckland in shock horror terms as resembling our modern bats or vampires but with its beak enlarged like the bill of a woodcock and armed with teeth like the snout of a crocodile. Its vertebrae, ribs, pelvis, legs and feet resembled those of a lizard. Its three anterior fingers terminated in long hooked claws like those of the forefinger of a bat and its body had a covering of scaly armour like that of an Iguana and in short it was a monster resembling nothing like that had ever been seen or heard of upon earth excepting the dragon of romance and heraldry.

Mantell visited Lyme Regis on one occasion, in 1832, meeting with Mary Anning, 'the Lioness of the place'.

> She was in a little dirty shop with hundreds of specimens piled around and in the greatest disorder. She, 'the presiding Deity', was a prim, pedantic, vinegar looking, thin female, shrewd and rather satirical in her conversation. She had no good specimens by her; but I purchased a few of the usual Lias fossils. Having heard of a fisherwoman who had a few fossils, we sought her out. She had a portion of the vertebral column of a Plesiosaurus dolicheideron and other bones which I bought.

Mantell had studied marine fossils from the chalk pits around Lewes, getting the workmen to search for anything unusual and rewarding them for their labours. However, he decided to look further afield into the outcrops of what he described as iron-sandstone in the Weald. Near Cuckfield, a staging post on the way to Brighton, in a wooded area known as Tilgate forest which had since become Cuckfield Park, were two stone quarries at Whiteman's Green abutting a stream by Taylor's Bridge. He made several excursions there either with his family, or, as on 26 September 1821:

> This morning at 10 o'clock I started for Cuckfield in a single horse chaise taking my young friend Rollo with me; we proceeded by Stonepond, through Ditchling and then followed the Brighton Road. The principal object of this excursion was to ascertain if possible the geological relations of the Cuckfield strata. We dined at the Talbot inn and then

> drove to the upper quarries; nearly 30 men were employed; we obtained a fine lumbar vertebra of a crocodile and a tooth of the same kind of animal.

The fossils obtained from this site were very different, consisting of animal bones and fossil plants, including a large trunk of a fossilised tropical palm. Other fossils also bore a close resemblance to tropical flora, such as Euphobia, a spruge found in the East Indies. The animals had dwelt among land vegetation. Their bones were enormous and without the subtle articulation of sea reptiles. Likewise he established that the palms were larger than even today's cyclads. He sought help from his staff and from others to collect and correlate living and fossil ferns, visiting Messrs Loddiges of Hackney who were suppliers of palm trees, cyclads such as Zamia, and other tropical exotica. The tropical vegetation among the stones of Tilgate forest were larger and more luxurious than any of the present era.

Mantell was the first to recognise the freshwater character of some of the beds of the Weald which had previously been considered to be of marine origin; finding alternations of mud and sand with an absence of pebbles and shingle thus demonstrating that part of the area was formed by sediments deposited by a river and not a sea. He discovered riverine shells, the bones of freshwater turtles and the teeth of crocodiles in addition to the vegetation. Later writers, such as M. A. Lower, depicted the primordial Weald yet more melodramatically:

> teeming with thick jungles, composed of graceful and arborescent ferns, conifers and cycadea; high sunny river banks where basked herbivorous Saurians, gigantic and terrible, while bat-like Pterodactyls rose on colossal pinions and fanned the sultry air.

In the frontispiece of *The Wonders of Geology*, Martin depicted the landscape diversified by hill and dale, by steams and torrents, the tributaries of its mighty river. It was clothed with arborescent ferns, palms and yuccas and inhabited by enormous reptiles. At the picture's centre the herbivorous Iguanodon is being attacked by a Megalosaurus and a crocodile, a foretaste of an image 'nature at war' that was later to send shock waves through Victorian society with the publications first of Robert Chambers' *Vestiges* and later Darwin's *Origin of Species*.

Tennyson in 1848 considered the best way to recall the lost world of monster lizards and giant ferns, was by standing beside a railway line at night. The flaming eyes of the locomotive's open fire-box, the thunder of its iron tread, and the vomiting aloft of sparks and smoke, turned the engine into a giant Ichthyosaurus.

Iguanodon. A restoration of an Iguanodon by Martin.

Baron Cuvier had established that chimeras did not exist. No animal could have the head of a lion, the body of a goat and the tail of a dragon. Each part of an animal had to fit the function of the whole. Thus a carnivore would have sharp teeth, a muscular body, clawed feet, and be capable of fast movement. Sharp teeth found alongside other bones in the rocks at Stonefield near Oxford and from the Tilgate beds led to the discovery of Megalosaurus, before Christmas 1821. The teeth, vertebrae, bones and other remains all indicated an animal of lizard type. Some fragments of a cylindrical bone, probably the femur, were found indicating that the animal was of gigantic size. Who initially discovered this creature is a matter of dispute. As Mantell wrote, 'vertebrate remains, so much mutilated and so partially distinct as to require more than Cuvierian sagacity to determine the prototypes'.

B. Van Valkenburgh (*Economist*, Feb. 2009) has compared the teeth of carnivores of the Pleistocene period with those of today. Modern mammals will occasionally break a tooth, especially if starving and trying to tear off the last piece of meat, i.e. when times are hard. Fractured teeth were more common in the past with an average frequency of 8 per cent in the Pleistocene compared with 2 per cent in modern times; and breakages among grey wolves and coyotes, living then and now, 4-5 per cent today and 7-10 per cent in the Pleistocene. Her conclusion was that there are fewer carnivores and less competition after the advent of man than before.

These early finds were inadequately delineated. Thus Robert Plot in 1677 described and illustrated what was probably the femur of a megalosaurus and Edward Lhwyd, whose book on English fossils (1699) included Ichthyosaurian remains, also depicted a Megalsosaurian tooth from Stonesfield. William Smith may well have seen similar bones, finding chunks of bone at Whiteman's Green in 1809, but could make nothing of them and never published his findings. Similarly Thomas Webster recalled years later that he had found large saurian bones near Hastings in 1812, John Kidd, the Oxford mineralogist, affirmed in 1815 that the Stonesfield slate included 'remains of one or more large quadrupeds', which he presumed to be mammalian.

In 1817 Mantell sent Etheldred Benett thirty-two specimens including a fragmented mandible of some marine animal from the chalk, with bones and teeth from the Weald. Buckland as curator of the Ashmolean Museum knew the Stonesfield exhibit, but had never examined it in detail. However, when Cuvier visited Oxford he was shown specimens from Stonesfield which he agreed were similar to crocodiles he had found at Honfleur. In June 1822 Mantell with the help of Lyell identified teeth and bones of crocodiles and other saurian animals of enormous magnitude which he published in May that year, thereby providing an anatomical description of a 'dinosaur'.

The first definitive paper was by Conybeare and Buckland in February 1824 describing the Megalosaurus or great lizard of Stonesfield. In April Conybeare published Mantell's related description of Megalosaurus together with his own, comparing it with Cuvier's findings at Honfleur. A few weeks later James Parkinson noticed Stonesfield material in his *Outlines of Orcytology* under the heading Megalosaurus, 40 feet long and 8 high. By comparing the thigh bone, which in the Oxford specimen was 2 feet 9 inches long, with that of an ordinary lizard Cuvier declared that the Stonesfield reptile must have been more than 49 feet long and as bulky as an elephant.

Following the papers of Conybeare and Buckland, Mantell provided a further less formal presentation, having brought with him enough specimens of Megalosaurus to convince Buckland that his own researches had been based on insufficient evidence. While crediting Buckland all too generously with the first discovery, despite the teeth, ribs, vertebrae and other bones which had been found in Tilgate forest, Mantell went on to suggest that it was an amphibian and not a marine creature; and probably lived in marshes. Later taking into account the flora found in the same strata and their size, he speculated that they walked on solid ground where plants and animals flourished in conditions analogous to that of tropical climates. Mantell's findings were initially hampered in getting past the referees and the publishing committee held up by the volume of transactions to enable the incorporation of Conybeare's and Buckland's hastily written paper. Buckland duly warned of the need for fairness, abandoned his efforts to use Mantell's illustrations and in characteristic style made generous references to Gideon Mantell, applauding his rich and highly valuable collection of fossils and acknowledging that the specimen of Megalosaurus in Sussex added to the knowledge of the animal.

Mantell's more remarkable finding was a large tooth which from the worn, smooth and oblique surface of the crown had evidently belonged to an herbivorous animal. This initially insignificant but unusual specimen led to the discovery of the Iguanodon, an enormous fossil herbivore. There are diverse recollections of the event, of which Mantell published several. He had travelled out to Cuckfield with his family on several occasions, but the first outing there had been a disaster with a thunderstorm. However, he was able to make a deal with the workman in charge of the quarry, Mr Leney, who regularly supplied him with interesting bones and other fossils. The earliest and most reliable source confirms he possessed six teeth from the quarry, several of which may have been found by his wife. He could have received the earliest package in June 1820 including an enormous bone, several vertebrae and some teeth (which he called Proteosaurus, Sir Everard Home's[5] name for the Ichthyosaurus). The worn-down incisor tooth of cuneiform shape had a broad, flattened, grinding surface, supported by thick enamel on one side and a marked ridge up the middle, which he presumed was the result of constant chewing. He described how 'the first specimen so entirely resembled the part of the incisor of a large mammal that I was much embarrassed to account for its presence in such ancient strata, in which according to all geological experience, no fossil remains of mammal would ever be discovered.'

On 21 June 1822 Mantell took the tooth and other Tilgate specimens to the Geological Society. Buckland, Conybeare and Clift were doubtful of his claims; but Dr Wollaston supported his opinion that the teeth might be from an unknown herbivore reptile. Lyell took the tooth with him to Paris to show Baron Cuvier. Mantell was later to describe Cuvier 'as a stout, square-built man, about the middle size, his face large and not so expressive as I had anticipated, a noble expanse of forehead; his eyes bright and penetrating'. On Saturday 28 June 1823 Lyell managed to get an invitation to Cuvier's Saturday evening soirée attended by

> the learned, and the talented of every nation, of every age and sex. All opinions were received; the more numerous the circle the more delighted was the master of the house to mingle in it, encouraging, amusing, welcoming everybody, paying the utmost respect to those really worthy of distinction. It was at once to see intellect in all its splendour; and

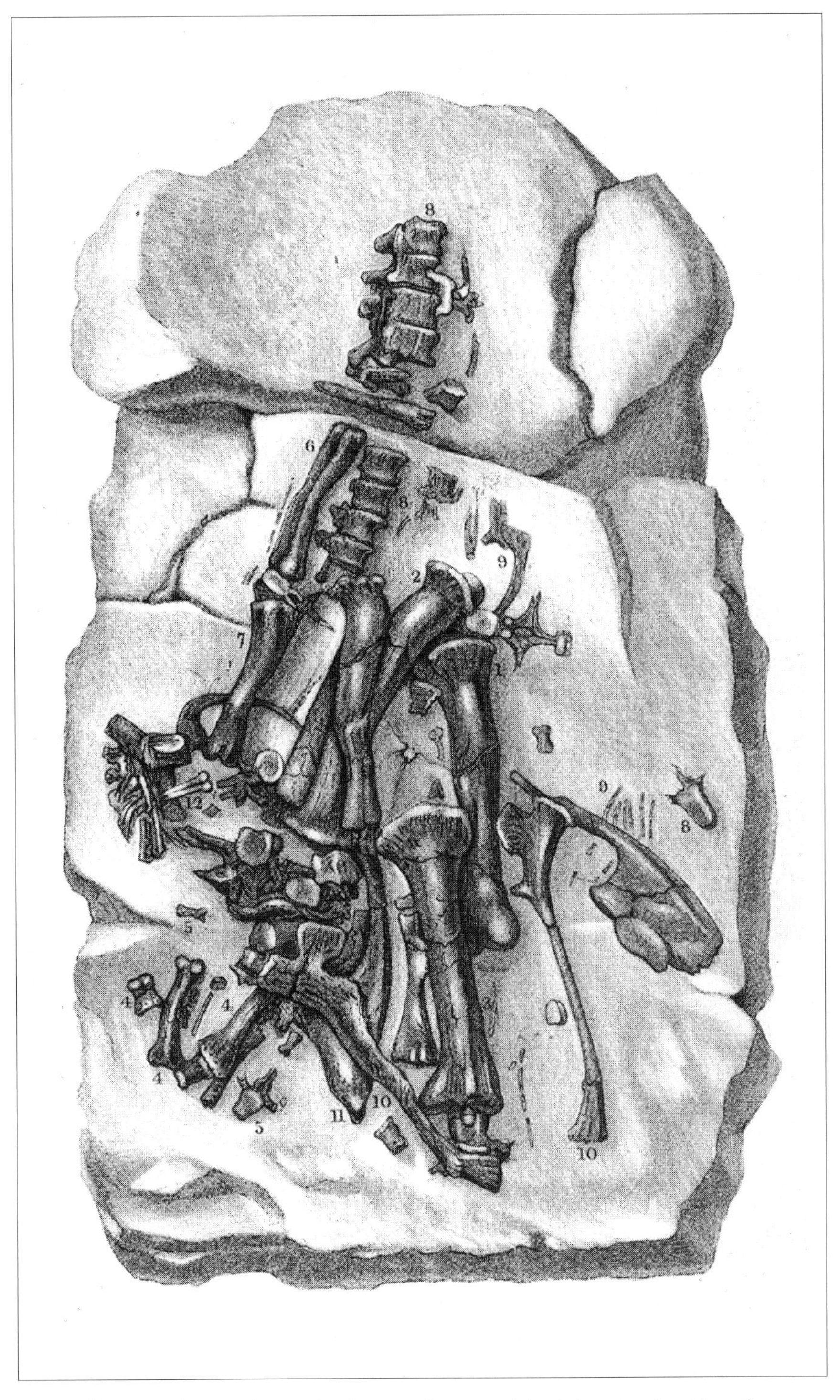

(Mantell's copy) Iguanodon and other saurians – original drawing by Mantell, published in *Philosophical Transactions of the Royal Society* (1841).

> the stranger was astonished to find himself conversing, without restraint, without ceremony, in the presence of the leading stars of Europe: princes, peers, diplomatists, and the worthy savant himself.

When first shown the tooth Cuvier dismissed it as a rhinoceros tooth. When Lyell persisted and presented some metacarpal bones these too were dismissed as from a species of hippopotamus. It is possible that Buckland had misinformed Cuvier that the Sussex rocks were recent. Mantell received a letter from Paris with the crushing verdict. Fortunately, as Lyell reported, on second thoughts the next day, he (Cuvier) believed it to be 'something quite different'.

Only after Mantell had sent Cuvier a graded series of deeply worn to nearly unworn teeth did he become fully convinced of his error:

> Ces dents me sont certainement inconnues; elles ne sont point d'un animal carnasssier, et cependent je crois qu'elles appartienment – Might we not have a new animal, an herbivorous reptile?

Mantell visited the Hunterian Museum of the Royal College of Surgeons to study the comparative anatomy of the teeth with the curator, William Clift. Clift came from a poor family in Devon but showed exceptional talent in drawing, so much so that the local gentry sent him as an apprentice to the surgeon John Hunter to assist him with his collection. The Royal College of Surgeons had been established in 1813 to assist the need for physiological instruction and training of surgeons. When Hunter died he left an important collection of natural history specimens to the college. The cataloguing of the collection was hampered by the fact that Sir Everard Home had removed and burnt most of the relevant manuscripts. Clift had slowly collated the specimens through comparison with other living creatures and some fossils. When Mantell produced the teeth, either a visitor from Barbados or Samuel Stutchbury, son of John Stutchbury the dealer, directed his attention to a 3½ foot skeleton of an iguana saying that there was a resemblance to the teeth of this smaller creature. The name 'Iguan-o-don' was suggested by the Rev. Conybeare:

> Your discovery of the analogy between the Iguana and the fossil teeth is very interesting, but the name you propose, Igunasaurian will hardly do, because it is equally applicable to the recent Iguana. Iguampoides, like the Iguana, or Iguanodon, having the tooth of the Iguana, would be better. The new herbivorous reptile was presumed to be of enormous size. Baron Cuvier agreed: "Just as the biggest living mammal happened

to be herbivorous, it seems quite possible that, when reptiles were the only kind of animal living on land, the largest among them also used to live off plants."

On 6 March 1824, Professor Buckland came from Oxford with Mr Lyell to inspect the Tilgate fossils. Mantell had met the professor at a meeting of the Geological Society about three weeks earlier and shown him some specimens of bones and vertebrae of the Megalosaurus from Tilgate forest. The next day he accompanied them to Cuckfield in the afternoon but it rained heavily limiting their visit.

> November 28th, during the last week there have been numerous applications from different persons respecting the new animal whose teeth I have discovered in the sandstone of Tilgate forest and which I have named Iguanodon. Have prepared parcels for Cuvier and Brongniant, M. Schlothesis and Baron Humboldt and two parcels of fossils for Dr Brown of Glasgow' December 7th Spent the evening at the Geological Society where I met Professor Buckland, Mr Lyell etc., etc. Had previously visited the College of Surgeons and had a long interview with Mr Clift and Mr Konig.

In the autumn of 1824, Georges Cuvier published his new edition of *Recherches sur les Ossements Fossiles*. He pointed out the debt he owed to Mantell. 'It is only since Mr Mantell sent me an entire series of teeth more or less used, that I am quite convinced.' With such public acknowledgement, Mantell was hastily admitted to the elite circles of London society and The Reverend W. Buckland wrote to him urging him to attend the next meeting of the Geological Society, and inviting him to dine in a tavern at St James Street.

On 24 March 1825 he was proposed for Fellowship of the Royal Society:

> Gideon Mantell Esq., of Lewes in Sussex, a gentleman well skilled in general science and particularly in geology, and known to the Society as the author of a paper on the fossil genus Iguanodon, being desirous of becoming a fellow – (signed) William Buckland Adam Sedgwick, William Conybeare, William Fitton, Davies Gilbert, Charles Konig, G.B. Greenough, John Abernethy, William Clift and others.

His paper on the Iguanodon was read before the Royal Society and Cuvier made handsome mention of Mantell in the second part of his fifth volume. Modern encyclopaedias define the Iguanodon as a heavily built, two-legged dinosaur up to 8m /26 ft in length, plant-eating, with a cropping horny beak present at the front of both jaws, and probably living in herds. They were known mainly from the Lower Cretaceous period of Europe (order Ornithischia).

Almost single-handedly and despite the virtually unanimous disapproval of his most esteemed colleagues, he had persuaded Baron Cuvier, the French authority, to change his mind, graciously recanting in public. On New Year's Day 1825 he supposedly completed and signed his most famous publication *Notice on the Iguanodon, a newly discovered fossil reptile from the sandstone of Tilgate forest in Sussex*, sending it by letter to Davies Gilbert who read it before the Royal Society on 10 February. It was an astonishing achievement for a relatively obscure provincial surgeon, still only thirty-four.

From the findings of other scattered bones near the teeth, Mantell was able to gauge some idea of the size of the Iguanodon. Some fragments of vertebrae were up to 5 inches long and a part of a rib measured 21 inches and the metatarsal bones of the foot were huge and chunky. Comparing these with those of an elephant, there seemed reason to suppose that they could have been of similar bulk, exceeding in magnitude every animal of the lizard tribe hitherto discovered. 'Like Frankenstein, I was struck with astonishment at the enormous monster which my investigations had called into existence.' Buckland said that such lizards were so colossal that an elephant would seem a shrimp by comparison.

On 6 May, Mantell received drawings and descriptions of some recently found fossils by W. H. Bensted a quarry owner of Maidstone in Kent. On 9 June, he travelled to Maidstone with W. D. Saull. Embedded in the Kentish rag, a marine deposit, was a fine specimen of an Iguanodon with two complete femurs, a tibia and fibula, a clavicle 28 inches long, large flattish bones, and vertebrae, mostly sacral, but also lumbar and possibly lower thoracic. In addition there were some ribs, remains of teeth and claw-like ungular bones; but no jaw. Mantell offered to buy the bones for £10 but Bensted insisted on £25. Moses Ricardo, a surgeon from a stock exchange family, and Horace (Horatio) Smith, a poet also involved in steam locomotives, with colleagues, put up the money and after working on the rock chiselling at it for three months, the Maidstone Iguanodon became a prize exhibit in Mantell's museum.

In 1834 a poem, probably by Horace Smith, appeared in the journal of the Amici Society (quoted by D. B. Norman, *Modern Geology*, 18: 225-45, 1993)

Our young Geologist who found
These monstrous Bones deep underground

And sent his parcel, not a light one
To his enlightened Friend at Brighton
Imagined, perhaps, like those who sent
The marbles of almighty Greece
Here, to some Antiquarian friend
They'd make a famous MANTEL-PIECE

In 1836 G. F. Richardson set the Iguanodon story in verse:

Tis indeed a world of wonder
Found within the earth and under
Forms as wild as fancy wishes
Monster lizards, stony fishes
Fragments of the lost amphibian
Here a femur – there a tibia-
Here the monster mammoth sleeping
Here the giant lizard creeping.

Over many years Mantell had accumulated hundreds of teeth of the Iguanodon without being able to prove how they formed the dentition embedded within the bony jaws of the reptile. An important inference resulting from a find by fishermen at Brook Bay in the Isle of Wight in 1847 of several vertebrae and teeth of an old Iguanodon suggested that, despite aging, the serrated margins of the teeth were not worn away due to continual renovation and replacement. In 1838 at Cuckfield, Mantell had uncovered part of the upper jaw (the maxilla) of an Iguanodon but without teeth in situ. However, on 25 March 1848 he received an unexpected package from a stranger, Captain Lambart Brickenden of Warmington, Sussex, the current proprietor of the quarries at Whiteman's Green where the original Iguanodon was discovered. Capt. Brickenden asked Mantell to describe the contents: a 21-inch-long specimen, very heavy and of an umber colour. To Mantell's delight it was the front part of a mandible (lower jaw) with seventeen teeth sockets and two tiny replacement teeth. As he wrote to Silliman: 'Here after 30 years search is an unequivocal portion of the dental origins of that marvellous reptile.' After acquainting Sedgwick and his other geological colleagues, he replied expressing great interest in the specimen and with Brickenden's gracious agreement gained possession of the specimen in exchange for a set of his publications.

Mantell took the mandible to the British Museum where, with Dr A. G. Melville, he compared the specimen with the jaws of other fossil and more recent reptiles. The rear, temporal portion was missing but the frontal, chin section, was intact and expanded in front with a scoop-shaped projection, with similarities to that of a mammalian sloth. On 18 May 1848 they were able to exhibit the specimen at the Royal Society having determined the positioning of the teeth in relation to the upper and lower jaws. A more technical analysis was read to the Royal Society on 25 May. The picture he presented was of a jaw as much as four feet in length with a grasping under-lip and a large, fleshy, prehensile tongue.

Professor Owen had previously assumed the head to be limited in size to two and a half feet at most; and on 31 May said that he would present his own specimen of an Iguanodon jaw, half the size of Mantell's, with five or six successional teeth in place but no mature ones. Owen's specimen belonged to G. B. Holmes of Horsham; and as a neighbour and fellow collector Brickenden had been able to see the jaw and provide Mantell with its particulars. It was from a very young Iguanodon a third the size of Mantell's. In December Mantell and Melville spent two hours looking over Holmes' Wealden fossils and, though not permitted to make either drawings or notes, they were able to make an analysis convincing themselves that many of Owen's reconstructions were based on imperfect information.

Mantell published the estimates of the size of the Iguanodon, based on the teeth, clavicle, femur and tibia, at between fifty-five and 100 feet long. From a later specimen found on the Isle of Wight he concluded that the front limbs were relatively small compared to the hind limbs upon which it walked. The recognition by Owen that the sacrum was composed of five fused vertebrae suggested a more moderate size; but in fact the Iguanodon, as opposed to some other dinosaurs, had six fused vertebrae. More Iguanodon bones were discovered while excavating a railway tunnel at Bletchingley and presented to the Geological Society along with specimens of the fossil plant, *Clathraria lyellii*.

HYLAEOSAURUS

In the course of the year 1830 the number of packages arriving at Lewes from Cuckfield with bones from the neighbouring quarries fell substantially. So much so that on 18 December Mantell drove to Cuckfield and endeavoured to obtain more fossils from the quarryman whom he had employed and paid to provide him with any finds of fossils over many years. 'The ungrateful scoundrel' refused point blank having found a customer on the spot willing to pay more. He returned distraught: 'Here then is the end to my hopes of discovering the jaw of the Iguanodon!'

However, approximately eighteen months later in July 1832, workmen at Whiteman's Green had been blasting at the hardest rock when they

Tilgate quarry, Whiteman's Green, Cuckfield, with Mantell in the foreground.

noticed petrified bone. Their explosion had caused so much damage that no amateur such as Mr Trotter who had been outbidding Mantell could deal with the material. A letter was despatched to Mantell informing him of the find. As Mantell reported to Professor Silliman:

> A mass of stone blown into fifty pieces or more was found full of bones. They reserved it for me, and with much difficulty and great labour, I have succeeded in uniting and clearing a slab of 4½ feet by 1½ feet exhibiting 12 vertebrae, eight in place, with many ribs, coracoid bones, omoplates, chevron bones, etc. and several of those curious dermal bones which support the scales.

Having collected the bones, Mantell washed, then successfully pieced together a block while leaving more than a wheel-barrowful of additional fragments. He concluded that it was the anterior portion of a fossil lizard, identifying part of the cranium, neck and dorsal vertebrae, ribs, dermal processes exterior to the skeleton, very long dermal spines each 15 inches long, coracoid bones extending from the scapula to the sternum, scapulae with glenoid cavities to articulate with the humerus, and other detached bones. He worked on the specimen until mid-October. This was the first armoured dinosaur which he labelled Hylaeosaurus armatus – armoured dinosaur of the Weald – and is presumed to be herbivorous.

He was unhappy with the reception of his paper on the Hylaeosaurus:

> none of our scientific journals have done me justice, not but they have praised me beyond my deserts, but they have not analysed the work nor noticed the wonderful Hylaeosaurus as he deserves to be noticed – the truth is Cuvier is no more and his *mantle* has not descended on any of our savants.

In 1835 he was awarded the Wollaston medal in absentia. Lyell, as President of the Geological Society, in his address said the award had been made for his long-continued labours with the comparative anatomy of fossils; especially for the discovery of two genera of fossil reptiles – Iguanodon and Hylaeosaurus.

Over the years Mantell was gradually able to sort out four other dinosaurs from bones gathered at Whiteman's Green. Among several bones whose characteristics were undetermined, all of them apparently saurian, Mantell noticed one that Buckland had assigned to a whale. Though it differed from the humerus of the only known plesiosaurus, Mantell supposed that the bone might belong to a new species within the same group. Several geologists had found similar bones of the Cetiosaurus

Hylaeosaurus (1832). The fragmented bones firmly embedded in rock.

which they named as 'Whale lizard' and believed to be a marine creature. The distinction from the Iguanodon was made by Mantell in Buckland's museum. Mantell was the first, and for a long time the only one, to realise that it was a land-living animal.

A humerus 54 inches long, similar to, but distinct from, that of an Iguanodon suggested an even larger animal about 84 feet in length which Mantell wished to call a Colossosaurus but was later persuaded to rename as Pelorosaurus from Homer's word for monster. He subsequently confirmed the new dinosaur with further discoveries of a scapula in 1851 and some splendid foreleg bones in 1852.

A third set of bones led to the discovery of the Regnosaurus northamptonii. The name was chosen based on the Sussex tribe of Regni. The lower jaw had originally been shown to the Royal Society in 1841, when it was believed to have come from a young Iguanodon. On re-examination, although there were similarities, it appeared to be sufficiently distinct to suggest a different species or a form of Stegosaur. No other examples of this fossil have come to light. A further specimen, again believed to be that of a young Iguanodon, was seen by Mantell when he visited James Scott Bowerbank, the distiller and microscope owner in Islington, to view some bones he had obtained from the Isle of Wight. In attempting to reconstruct the badly crushed vertebrae and other bones Mantell discovered what appeared to be a very much smaller and lightly built Iguanodon that lacked the typical grasping thumb-spike. A plate illustration was added as an addendum to his publication on the Iguanodon's dentition. More recently in 1972 the bones were re-assigned by P. M. Galton to the possibly warm-blooded dinosaur, Hypsilophodon.

In conclusion, Mantell could boast that he had primary discovery of the Iguanodon and Hylaeosaurus and was involved in the correct understanding of the characteristics of Cetiosaurus, Pelorosaurus, Ichthyosaurus, Plesiosaurus, Megalosaurus, and Hypsilophodon.

THE AGE OF REPTILES

Mantell declared that before the existence of the human race the earth was peopled by oviparous quadrupeds of most appalling magnitude and that the reptiles were the Lords of Creation. In *The Wonders of Geology* (1838) he described the County of the Iguanodon:

> A beautiful country of vast extent, diversified by hill and dale, with its rivulets, streams, and mighty rivers, flowing through fertile plains. Groves of palms and ferns, and forests of coniferous trees, clothed its surface, and I saw monsters of the reptile tribe basking on the banks of its rivers and roaming through its forests, while in its ferns and marshes, were sporting thousands of crocodiles. Winged reptiles of strange forms shared with birds in the dominion of the air, and the waters teemed with fish and crustacean.

The Age of Reptiles could be defined in terms of the sequence of rocks immediately after the formation of coal measures, i.e. the Carboniferous period. The rocks of the Mesozoic era, 145-65 million years ago, are now divided into three periods: the Triassic saw some lizards and crocodiles; the Jurassic when the earth teemed with oviparous quadrupeds inhabiting in equal measure the sea and dry land; and the Cretaceous including the younger Tilgate rocks in the Weald of Sussex. With the chalk the age of reptiles may be said to terminate.

Together with the melodramatic prints produced by John Martin, these descriptions of saurian reptilian inspired many authors. The younger Gradgrinds in Dickens' *Hard Times* had a nursery filled with fossils which formed the basis of their early education. Mary Shelley in her gothic novel written as part of a story writing competition, in which her husband, Percy Bysshe Shelley and Lord Byron took part, had Frankenstein's monster roaming the globe before escaping to the Arctic. Mary Shelley later became a neighbour, patient and friend of Mantell in London.

PIONEERS

How should one describe Gideon Mantell? John Gribbin in his *History of Western Science* (1543-2001) describes him as a surgeon and amateur (but very good) geologist who also discovered several types of dinosaur. Robert Bakewell whose *Introduction to Geology* was highly popular and inspired many to enter the field was likewise described as an amateur. So could many others: William Smith the surveyor pioneered the discovery of strata, James Hutton qualified as a barrister and then as a doctor but took up geology. Alfred Russell Wallace after attempting various trades became a naturalist. The same could be said of Charles Darwin. Were they all amateurs? Do we classify Baron Georges Cuvier, Charles Lyell, and the Rev. William Buckland as professionals because they held posts specifically in geology or comparative anatomy? Buckland later became the Dean of Westminster. Lyell depended on his family fortune and estate in Scotland. James Parkinson though working as a surgeon made his name from his collection of fossils before achieving fame as a physician. Essentially they were all pioneers, attempting to establish a scientific basis for geology or palaeontology.

Mantell made his money while in Lewes as a surgeon-practitioner. In Brighton, where his practice was reduced to a trickle of patients except briefly during an epidemic, he existed on the benevolence of Lord Egremont and later had to sell his collection. In London he practised medicine once again, despite ill-health. He pursued geology with an overriding ambition, while developing other skills as a true polymath. He collected fossils and antiquities from an early age, activating quarrymen to look for and set aside fossils which he bought, catalogued, and examined minutely. Parcels from Cuckfield, Wiltshire, Cornwall and Scotland, and later from abroad, arrived at the Lewes wagon office. He examined the chalk downs and the marine fossils they contained at the same time recognising the way the Weald had come into being. He discovered the land or swamp fossils of the ironstone strata of the Weald, finding the fauna as well as the early dinosaurs. No dilettante, it took Mantell thirty years to convince his professional colleagues that the tooth from Tilgate Forest was that of an herbivorous dinosaur. He surveyed the Isle of Wight, developed a museum collection which he was later forced to sell to the British Museum, participated in the discussions of many learned societies where he presented his findings, and became a noted lecturer in London, Brighton and elsewhere.

Compared to the other geological pioneers he made relatively few mistakes. In the field of comparative anatomy he could be charged with failure to identify the horned structure which was probably attached to the fore-limbs of the Iguanodon and used for grasping; he was not the first to recognise that the sacrum was fused, and probably overestimated the size of the Iguanodon. But Richard Owen, as Professor of Comparative Anatomy, was slow to realise the bipedal stance of the Iguanodon, and made other errors which he failed to correct.

In many respects Mantell was ahead of his time. Although a member of the Geology Society, its committee refused to accept his view that the Weald formed part of a freshwater lake and delayed publication for over three years. Not until Dr Fitton, its secretary, visited the Weald with Mantell, and recognised the undulating furrows or ripples in the sandstone did he agree that the rocks might have been formed in a low-lying flood plane or along the edge of a lake or river where sandbanks had accumulated.

Over and over again Gideon Mantell made intelligent speculations which took years to prove correct. Such people create interest but are treated with caution; whereas the conventional follower is readily taken into the establishment. Browning's declaration 'that a man's reach should exceed his grasp, or what's a heaven for!' could be a step too far.

Mantell was the first and for a long time the only person to identify the Cetiosaurus (assumed to be a whale lizard) as a land-living dinosaur. When on 29 October 1835 he encountered a chimpanzee at London Zoo he stated that it was the 'nearest approach to the physical structure of man I have ever seen'. He agreed with Lyell that the earth had formed over a long period in stages and spoke of the Age of Reptiles before that of mammals. Once again he came up against the stifling authoritarian beliefs of the Victorian scientists who ordered him to omit the last paragraph of his innovative paper on the soft parts of forminifera in 1841 in which he predicted that traces of microscopic life would someday be found in so-called Azoic (lifeless) rocks underlying the Cambrian:

> I therefore submit ... that in the present state of our knowledge of the earth's physical history, as derived from palaeontological evidence, the period when organic creation commenced must still be regarded as one of those hidden mysteries of nature from which science has not yet withdrawn the veil.

LEWES

Mantell began his diary on 1 January 1819 and continued it until his demise, beginning by recording the events of the previous year. One reason for attempting to keep a journal is that he repeatedly forgot dates. In his first journal entry he misdates his barrow opening in 1811 by four months. Similarly he misdated Ellen Mary's birthday by a day. Dates in letters do not accord with journal dates and he was to misdate the essay in which he first described the Iguanodon.

Like all diaries it is necessarily egocentric, but from it one learns of his relationship with his family. His brother, Thomas Austen Mantell, was fifteen years older than Gideon and became head of the family on their father's death in 1807, when Gideon was just seventeen. He always refers to Thomas as 'my brother', and Thomas would take the lead in political matters but ask Gideon to write the resolutions. He continued to take an interest in Gideon's welfare, helping him to find a house in London and calling on him when he was ill in the years after his accident. His other elder brother, Samuel, he refers to by name, noting that Sir G. H. Crewe and his lady paid a visit to their niece – my brother Samuel's wife. His own wife, Mary Ann, he always refers to as Mrs Mantell but gives the Christian names of members of her family. His younger brother, Joshua is always referred to by name; thus

> my brother Joshua delivered a lecture at the Star rooms on the nature and properties of the gases illustrated by experiments – upwards of 200 persons were present; he acquitted himself remarkably well.

Until he became the principal in the practice he was dependent on his brother Thomas in various matters; thus the family would travel to Cuckfield in his brother's chaise with Gideon following on horseback, and

Mrs Mantell (1795-1869) by W. M. Woodhouse, photograph by R. J. Glaister.

rather than a sudden burst of opulence and extravagance as suggested by Dr Gordon Hake, the diary relates the gradual acquisition of paintings, engravings, and painting the crests on the marble table.[1] In 1821 he purchased a fine painting from Lord Hampden on the subject of the allegorical decline of the consular government of Rome for 10 guineas. He was highly pleased with the purchase of a case of Napoleonic medals, ten in number, consisting of the chef d'oeuvre of Andrieu, enclosed in an elegant morocco bookcase. A rocking horse presented to his daughter Ellen by Mr Baldock cost 5-6 guineas.

His wife initially shared his interest in fossils and sketching. She did illustrations and plates for him. They would go together to draw the landscape using a camera lucida. Mary was supposed to have found the first teeth of the Iguanodon. Women generally avoided politics and took little part in religious discussion but were not excluded from public societies. They enjoyed increasingly complex social lives often predominating in assemblies, plays and concerts; all of which he was happy to attend when he could with his wife.

He was very much concerned with illnesses within the family. He sat up part of the night with his wife who had asthma and bloody expectoration. He was obliged to bleed and blister. He sat up with his son, Walter, who had a chest complaint, and at two o'clock applied two leeches to the chest which bled profusely and greatly relieved him. He was apprehensive when his daughter Ellen was far from well with a mesenteric affection. Despite his heavy commitments, he frequently travelled with his family and visited his own and his wife's relations in London. Mrs Mantell's pregnancies seem to have been uneventful but her health in other respects was not good. In 1826 she was very gravely ill and confined to bed for a month, afterwards she went to Brighton to convalesce, returning after two months. On another occasion when unwell she went to her family in London and when expected to return, she had to delay her departure because of further ill-health. We are not told the cause of her health problems, though she did have chest troubles earlier on. While accepting their genuine nature, there is an increasing suspicion of psychological overlay. The first hint of domestic unhappiness is to be found in his journal in March 1824.

His working day was mostly spent seeing patients, both in Lewes and riding out to the villages, commenting as he goes on the weather, the flowers, crops and harvest and the state of the people. He mentions his assistants, Rollo and Llashmer, when he visits the quarries and gives details of their finding and growing exotic ferns and other plants, but gives no indication of his instruction to them on medical matters. During the day he would receive packets of fossil finds from the flint quarries around Lewes, Jenner's quarry, Malling Hill and Street pits, Southerham pit and occasionally from Cuckfield. Interested visitors would be welcomed to see his collections. In the evening he would attend meetings or dine with friends – often these were cancelled or interrupted with medical emergencies or midwifery calls – or he would sit up playing chess with Mrs Mantell; an event which became so rare as to require special mention. Late into the night he would sort, clean, examine, sketch and write about his fossil finds; ending with entering the events into his diary. Bills are sent out every January and there were particular dates, such as his birthday which were causes for introspection.

Certain hobbies were shared with the wider family, notably astronomy; 'The planet Ceres was visible in the telescope (owned by his brother, Thomas) last evening'. Ceres was the first asteroid to be recognised on the opening day of the nineteenth century. It has a diameter of 933 km, is nearly spherical and rotates every nine hours. Its discovery was made by an Italian astronomer and named after the goddess of fertility. William Herschel later coined the term asteroid. Four months later, they observed a comet 'very brilliant, the nucleus very distinct'. The next evening the comet was but slightly perceptible being partly obscured by the density of the atmosphere. The following year, in September they went to Balsdean, on a ridge of the Downs above Kingston, to view the eclipse of the sun

Southerham chalk pit.

from half past twelve to three o'clock. Its greatest obscuration was at thirty minutes before two, when it had a small crescent, which gave more light than the astronomers expected, the darkness much overrated. Three days later he accompanied Mrs Mantell 'to my brother's to view Jupiter through his telescope'. The moons (at least three of them) were distinctly seen and also the belts. The next day he rode with Mrs Mantell over the downs to Balsdean. The four moons of Jupiter were beautifully apparent. He also went with Mrs Mantell to Mr Woollgar's to observe the moon through his new telescope:

> The moon was near the full, and the shadows projected by the eminences on her surface were less distinct than had been anticipated. Jupiter with two of his satellites was beautifully distinct and also Saturn with his ring, and one satellite.

He was interested in new inventions such as the Pedestrian Accelerator Hobby-horse or velocipede and De Loutherbourg's Eidophusikon – a moving panoramic peep-show analogous to his stage scenery and melodramatic landscapes.

He watched, when walking with Mrs Mantell, Charles Green's (1785-1870) ascent in a balloon from the gasworks at Brighton, viewing it from the top of Keere Street. The balloon fell into the sea off Cuckmere and was picked up by a packet; the aeronaut was almost drowned. He may also have seen Mr Green ascend from Lewes with a passenger in September 1828. A Baxter print in the Barbican Museum shows the balloon with the Downs in the background. Mantell made an ascent to a considerable height along with old Mr Lee, Warren Lee's father and for fifty years editor of the local paper. 'He was a man of middle size, rather corpulent, with shaggy hair which would have been grey had he not kept it a mahogany colour by Atkins Tyrian dye; a remarkable prominent but thin nose, sharp grey eyes, and a peculiarity of physiognomy hard to describe'. Green made 526 ascents in his balloon filled with carburetted hydrogen gas, most famously travelling from Vauxhall to Nassau in Germany. He lived until eighty-five years of age.

Mantell writes excitingly of other adventures:

> May 9th 1825 I rose at three, left home at 4 o'clock with Mrs Mantell in my chaise; the morning remarkably fine. Drove to Crowboro' where we baited [took refreshment] soon after seven, from thence to the high rocks where we stopt, and rambled among the immense masses of sandstone, which were highly interesting; proceeded to Tunbridge Wells, where we breakfasted soon after ten. Drove to Penshurst Place, which I had not

> visited since 1809. Although much of the buildings was in ruins, and the paintings and statues cruelly spoiled by neglect, yet there was very much to interest me; the birthplace of Sir Phillip Sydney and of Algernon, could not but be interesting to every one; we visited the oak in the park (that was planted on the day Sir Phillip Sydney was born) and my name and my friend Tilney's which we cut on the inside in 1809 were still fresh and legible. Went into the church, visited the Sydney chancel. Drove to Tunbridge, viewed the castle and returned to the Wells; from whence we departed at six, and arrived home at ten o'clock, after one of the most agreeable rambles I ever had.
>
> June 25th. Mrs Mantell, Ellen and myself went to Steyning in a chaise. Viewed the church at New Shoreham, which is one of the finest architectural remains of the mixed Saxon and Norman styles in the whole County. A man was repairing some of the stonework of the windows etc. but I am fearful without sufficient regard to the original. At Steyning we visited the church, which contains some of the richest Saxon arches I ever beheld. Discovered the 'firestone' towards the North East of the town; obtained a few fossils from the chalk quarries on the hill. These consisted principally of the supposed 'Juli' of the Larch, and scales of fishes. In a cornfield I discovered a plant of Sinapis arvensis, having a most singularly looking flower. I have sent it to Mr Don, Librarian of the Linnaean Society. On our return the wind was very high, and the tide at the full; the appearance of the harbour was grand in the extreme. Several vessels were coming in, and the wind being contrary, they were obliged to tack very frequently.

Mantell saw himself as an active member of the community, viewing political events with a fervour shared with many of the townspeople; so much so that he records receiving a book on politics from an unknown person. He describes the dreadful catastrophe that occurred in Manchester (the Peterloo Massacre) in Saint Peter's Fields. A public meeting was convened for the purpose of petitioning the Regent; an orderly and unarmed crowd of about 60,000 men women and children were permitted to assemble; but then the magistrates, stricken with alarm at the sight of so great a multitude, sent in the yeomanry to arrest the speaker, the Radical Hunt, after the meeting had been in progress for some time. When the horsemen pushing their way through the throng were shouted at and hustled, the magistrates most unwarrantably, and illegally, ordered a troop of 60 cavalrymen, brandishing sabres, to charge and disperse them – they galloped into the immense multitude with their sabres drawn and made a 'most promiscuous slaughter', killing eleven, and severely injuring 500 people. The action of the government in sending the magistrates their thanks without waiting to make an inquiry, aroused an outburst of indignation among many respectable citizens who would never have dreamt of going to the Peterloo meeting. Everyone condemned this infamous measure – subscriptions are entered into in London, Brighton, etc. for the sufferers and for legal prosecution of the magistrates.

Later that month a requisition signed by upwards of thirty of the inhabitant householders of the Boro' was presented to the constables requesting them to appoint a meeting of the inhabitants to express their opinion on the proceedings at Manchester. The senior constable refused to comply with the requisition, the other constable, Mr William Stuard, immediately assented.

> At the request of my brother who intends to take an active part in the approaching meeting, I have this day composed a set of Resolutions, and an address to his Royal Highness, the Prince Regent...

The meeting in County Hall was conducted with great decorum and propriety. Resolutions deprecating the conduct of the authorities on 15 August, and an address to the Regent imploring him to dismiss his ministers, and to institute an enquiry into the late fatal events, were unanimously agreed to. His dislike of the Regent led to his gloating when a statue of the Prince was demolished in Brighton.

Following the death of King George III, the Regent was proclaimed King George IV: 'Scarcely an individual felt interested in the accession of the new sovereign who is very unpopular.'

There was a problem in that Lewes supported the Whigs while the two county members were both Tories, 'and voted for the infamous bills now commonly known by the term Desperate Measures'. Mantell met Mr Cavendish at the Crown Inn and afterwards accompanied him in his canvas through the Town. Two days later an immense number of freeholders had passed through the town on their way to Chichester, the greater part supported the interests of Burrell and Curteis; or rather were represented by the Earl of Abergavenny. Not to be outdone in defiance of the Tory clique, four days later he set off in a coach and four, for Chichester accompanied by Ebenezer Johnston and his late partner Mr Moore.

> Met by a large body of friends of Mr Cavendish who conducted us to the Poll. Six men, fantastically dressed in the colours of Mr Cavendish, walked on each side of our carriage; and a band of musicians with banners, etc. preceded us. Caesar in the midst of his triumphs would not have been received with greater enthusiasm. At the close of this day's

> poll Mr Cavendish had gained 53 upon Mr Curteis. Obtained lodging for Mr Moore who is very infirm.

The next morning they rose at 6.00 a.m. and took part in a promenade through the city but later found out that Mr Cavendish had declined to continue the contest.

His next political concern was the arrival of King George's queen from the Continent. In April 1795 the then Prince George had been persuaded to marry his poorly educated first cousin, Caroline of Brunswick. Bride and groom met for the first time at the Duke of Cumberland's house in Cleveland Row. When the Prince entered the room Caroline went down on her knees to him. The Prince raised her and embraced her, said barely one word and retired, calling for a glass of brandy. On the morning of his marriage the Prince of Wales told his brother the Duke of Clarence, 'William, tell Mrs Fitzherbert she is the only woman I shall ever love' and fortified himself with so much brandy that he was noticeably inebriated during the service. Soon after the birth of their only child, a daughter christened Charlotte Augusta, the Prince sent his wife a message through his Chamberlain that he intended never to share the same room with her again.

Even before the death of Princess Charlotte in childbirth they went their separate ways. Both the Prince Regent and Caroline took several lovers and she lived a debauched life mainly in Italy. As the rumours mounted, Jane Austen gave her opinion that 'she may be bad, but he made her so, and he is worse'. In the public imagination the Peterloo massacre, the Cato conspiracy and the treatment of the Queen Consort were used to attack the king and the establishment. Crowds greeted the queen when she returned against the king's wishes to Britain. William Cobbett and his daughter Anne were among the crowds, and waved a bough of laurel, as they watched the queen pass by in a miserable half-broken-down carriage covered with dust, followed by a post-chaise and a calash. More cynically, Percy Bysshe Shelley said that 'making her Sacred Majesty into a heroine demonstrated the generous gullibility of the English'.

The king had had her name removed from the liturgy and was now able to pressurise parliament to allow him to seek a divorce. In the public imagination she was considered as greatly injured, and in consequence her health was drunk at all public festivities with great enthusiasm. With his secret marriage to Mrs Fitzherbert the king was already considered a bigamist; and against the Constitution had married a Catholic. The conduct of Lord Sidmouth in employing spies to instigate Thistlewood etc. to the Cato-Street conspiracy, was universally spoken of in terms of detestation and an abhorrence. The queen's conduct was much praised and the king almost universally blamed. The Cato-Street Plot was a conspiracy in February 1820 formulated by Arthur Thistlewood and fellow radicals to blow up the British Tory Cabinet as it attended a dinner at the house of the Earl of Harrowby. The plot was infiltrated by a government spy and agent provocateur called George Edwards, and the leaders were arrested, hanged and decapitated.[2] Gideon Mantell and many others feared that the infamous bill divorcing the queen would be carried through the House of Lords by the corrupt influence of ministers. At County Hall the address to the queen, very strong and pithy, was passed without a dissenting voice. Following which the constables and aldermen went to support the queen. 'Mr Figg describes her Majesty as an interesting woman, not handsome, but very fascinating in her manners.'

In November the second reading of the infamous Bill of Pains and Penalties against the queen was introduced to the House of Lords and conducted there by the examination of the evidence as in a Court of Justice. On 21 August Mantell wrote in his journal:

> The mock trial of the Queen is going on; the Italian witnesses are of the most infamous description; they have evidently been bribed or repeatedly catechised in their evidence.

The chief witness, Majocchi, broke under cross-examination repeating notoriously the phrase 'Non mi ricordo' (I don't remember). Even so the bill was carried by a majority of 28, contrary to the wishes of nearly the whole nation 'who are more than ever convinced of her innocence'.

The *Lewes Journal*, edited by Mr Lee, was publically burnt before Mr Lee's door because it contained a letter against the queen supposed to be written by the Earl of Sheffield.

> The next day, however, I was awoke at 5 o'clock this morning by the ringing of the Church bells, the tolling of Old Gabriel, and the rejoicing of the people in the streets, in consequence of the bill against the Queen having been thrown out of the House of Lords. All business is at standstill, everyone is rejoicing; the poorer classes decorated themselves with laurels: the genteel folk wore red roses.

Mantell's personal involvement took the form of a transparency displayed in his drawing room window painted by his brother-in-law. The centrepiece was Caroline regina. The imperial crown surmounted a garter on which was inscribed 'Dieu potege le Droit'. At the base was a Hercules club on which was written 'vox populi', and a dead serpent twined round a club had the word 'Calumny' written upon it. The whole was surmounted by a wreath of oak and laurel.

When the queen died, Mantell commented that

> the Queen was interred without the funeral service being performed; such despicable revenge will only recoil upon the heads of her enemies. It is astonishing how universally the conduct of ministers in this affair is deprecated.

In a diary entry he rails against the abduction and subsequent treatment of Napoleon.

> This truly great man bore his fate with the greatest fortitude and resignation, and although suffering for several years from a most excruciating disease; and daily harassed by the vile minions which our wretched ministers have placed over him as jailors, he bore it without complaint. The hireling journals are loading his memory with infamy; poor creatures – they are perfectly harmless – the name of Napoleon will be venerated by prosperity.

*

After James Moore retired Mantell ran a single-handed medical practice from his home for several years before taking an assistant. He had two apprentices. George Rollo arrived having served four years' apprenticeship before his master went bankrupt. He was the nephew of Dr Rollo of Woolwich Hospital who was the author of a treatise on diabetes. George Rollo punctured his lung while dissecting in London and later died from an infected finger from an injury in the dissecting room. Mantell sacked a second apprentice, Lashmar, for improper conduct; repaying his father £59 as a proportionate part of the premium received from him. A third apprentice, Lawrence Alford, left after four months.

The practice involved frequent visits on horseback to the neighbouring villages. Once in consequence of the illness of his horse he was obliged to visit his country patients on foot, seeing eight in Ringmer, further visits to patients in the hospital, applying dressings to a patient with burns, walking to Norlington and returning home at eleven o'clock. Progression on horseback was rarely faster than four miles per hour with difficulties encountered in all weathers – snow, ice, mist, darkness and thunderstorms. Once his horse was struck by lightning and its ears became phosphorescent.

Going to a patient at Ringmer Lodge, it rained in torrents, was exceedingly dark, and he could not distinguish the footpath except when flashes of light threw a sudden illumination on it. With great difficulty he found his way to the house and to his great joy the accouchement had taken place, as he was drenched with rain and hail. He resolved to return home immediately, and therefore notwithstanding the rain, which continued to pour down most violently and the lightning which was frequent and very vivid, he mounted his horse and endeavoured to find his way out of the enclosure which was separated from the turnpike road by a hedge and gate. After considerable trouble he arrived at the latter and endeavoured to ride through it, but although the gate was partly open, his horse refused to proceed; doubtful of being right, he hesitated whether to urge him to go forward but by a vivid flash of lightning he perceived the gate very distinctly and used every exertion to induce his horse to press on. All this time it hailed and rained in torrents, and in fact blew a complete hurricane.

At that moment, a remarkably vivid flash of lightning seemed to strike the ground near him, and no sooner had it vanished, when to his terror and astonishment, both the ears of his horse presented a luminous appearance, which he could compare to nothing more like than the lambent light produced by phosphorus or rotten wood. The edges of both ears were more luminous than the rest. Resolved to satisfy himself that there was no optical illusion, he leant forward with his hand (which was covered with a wet leather glove), stroked both ears two or three times, but the luminous appearance still remained. The horse started back and became so unmanageable that at last he galloped into the field and after a short while Mantell conducted him back to the farmhouse and called for assistance. A man brought out a lantern and conducted him through the gate into the turnpike road. The light was still on the ears of the horse while the man was getting the lantern, but when he came out, the light of the candle prevented ascertaining how it had disappeared, for when the man left no vestige remained, and it was so dark that Mantell could not distinguish the head of the animal. The storm gradually vanished, and before he got home it was starlight and serene.

Fortunately the horse recovered. On another occasion he was thrown from his horse, narrowly escaping being killed, and hurting his shoulder exceedingly. He was obliged to apply leeches to the part. The only other malady recorded at that time apart from colds and fatigue was a bout of giddiness of the head one evening. He applied fifteen leeches to the temples from which he experienced considerable relief. (This was probably a migrainous equivalent). He regularly bought leeches for his practice from London in batches of a hundred, priced at ten shillings.

In his practice he had to cope with an outbreak of typhus, seeing sixty patients in one day. His mother-in-law and the servant were both infected but recovered. His treatment of typhus followed that advocated by Dr

John Armstrong of the London Fever Institution – a physician he knew personally and admired, and who sent him the third edition of his work on Typhus Fever. The servant girl was given a warm bath and an 'affusion' of cold salt water which reduced the progress of the fever. He wrote to Mr Pearson for his opinion on a case of gangrene at Piddinghoe. He continued to attend Mr Moore and when a patient was severely burnt, despite the poor prognosis, he applied fresh dressings taking an hour each day until she died three weeks later. During epidemics of typhoid, typhus, smallpox, malaria and measles he declared: 'Our house is like a public office from the continued ingress and egress of persons sending for their medicines.' Smallpox broke out in the same month with fourteen deaths despite an earlier vaccination campaign.

MEDICAL PRACTICE

Standard treatments such as Mantell would have prescribed consisted of blood letting, cathartics, emetics, sudorifics, and febrifuges (for fever). Those with consumption (phthisis) were given a low diet avoiding animal fats and wine, had frequent blood letting and were told to avoid the cold, or excessive exercise or emotion. When pregnant, patients were advised on diet, exercise, regulating the bowels, and bathing. Practitioners tended to have their individual panaceas, inevitably highly elaborated and over-prescribed. Of the specific therapies only four would pass scrutiny today – quinine, digitalis, laudanum (tincture of opium) and after Lind (1757) fresh fruit and vegetables. Ether and chloroform became available in the latter part of his career. Although he experimented with them to ease his spasms, there is no evidence that he used them to treat his patients, but he did witness operations involving their use. Two forms of therapy were almost universally applied: bleeding and the application of leeches.

Blood letting was also used in all eventualities, such as vomiting or where a course of action was unclear. Such treatments depended on their psychological effect, partially treating hysteria and gout in the rich, and accidents with bruising in the poor. Bleeding may have helped where gout was associated with myelo-proliferative diseases such as polycythemia. It could be performed by a simple lancet or by scarification. Leeches not only inflicted less pain but removed a more predictable quantity of blood. They could be applied to anatomically awkward areas, e.g. tonsils, haemorrhoids, and the cervix, by means of a leech glass, but could also migrate to disappear inside the rectum, etc. Cupping involved the application of a heated glass vessel to an area of skin creating a partial vacuum, drawing blood or pus to the surface (in effect acting as an artificial leech).

Distressing accidents occurred at the workplace, and there are many examples treated by Mantell.

> An iron bar falling on an apprentice of Rider the Blacksmith, dreadfully lacerated the hand. It was necessary to remove the forefinger and thumb at the wrist joint. At Chailey Mill a poor boy got his clothes entangled by an upright post that was revolving; the lad was whirled round with great velocity; and his legs were dashed against a beam; the consequence was a separation of the left tibia from its articulation with the knee joint and a laceration of the whole muscle of that leg; on the right side the femur was knocked away from the epiphysis and projected between the muscles of the ham; but in both instances the joint of the knees remained perfect and uninjured. It was considered absolutely necessary to amputate the left leg above the knee, the right femur was reduced and kept in place by appropriate compression; but the constitutional shock was so great that the boy died the next day.

Mantell trephined a patient from Barcombe with a fractured skull but did not expect the man to recover.

Accidents due to wagons running over people were frequent. 'One of the Artillery drivers was nearly killed by a wagon passing over him. Brought to my house almost dead and sent him to hospital.' He examined the head of a boy who 'was drove over by a cart about ten days since'. The skull was fractured and the meninges lacerated with a large coagulum pressing on the brain and tetanus occurring a few days before death. There were other head injuries, fractured thighs, and the need to amputate thumbs, fingers and even forearms. For some of the more difficult operations he would assist Thomas Hodson (1762-1841) who had trained in London and gained a reputation for lithotomy (removing stone from the bladder). He had moved to Lewes in 1797, buying Newcastle House. Together they removed a fungus hematodes from the side of a poor man who had had it for years, operated on a tumour of the breast and on an inguinal hernia. The latter patient was a labourer in the employ of Lord Gage. His Lordship had sat up with the man nearly all night. He was called in consultation by Mr Hodson for a gunshot wound to the wrist causing profuse haemorrhage. He assisted him in the amputation of the thigh of a poor man who had long suffered from a disease of the knee joint, and Mr Hodson assisted him in deftly amputating a mangled forearm. Gideon Mantell was also successful in performing cataract operations, dilating the iris with belladonna, probing with a needle, and dividing the clouded lens.

He sat up all night with a lady following an abortion. He was required to perform autopsies on patients dying of consumption, on a lady with a large ovarian cyst containing six gallons of fluid and on a young girl dying in childbirth with fluid in the abdomen and bleeding into the pericardium; the uterus and its appendages were perfectly healthy.

He gained fame medically in 1826 when Daniel Leney was hanged, a mere boy lodging with Hannah Russell of Burwash. The husband and the lodger were in the act of stealing corn from a neighbouring farm in the dead of night, when Benjamin Russell who was carrying a heavy sackful of corn suddenly collapsed and died. Gossip among the villagers of Burwash soon produced circumstantial evidence pointing to Mrs Russell and the lodger as responsible for his death. One witness said that Mrs Russell had been seen purchasing a packet labelled 'poison' at the village shop, while another said that Mrs Russell had been seen spreading a white powder on a slice of bread which she said was for the mice. A post mortem was performed and the coroner was told that a drachm of arsenic had been found in the stomach of the deceased. Both were convicted in Lewes of poisoning her husband after arsenic was diagnosed by a witless surgeon whose expertise Baron Graham warmly commended. Gideon Mantell claimed that the arsenical test used by the medical witness was inconclusive, a claim which was upheld by Professor Brande and four eminent doctors. Furthermore Mantell had evidence that the deceased suffered from angina pectoris. Legal arguments over her marital status delayed Hannah's execution, enabling Mantell by the exertion of great interest and solicitations in addition to his scientific efforts to secure the woman's pardon long enough for Gideon Mantell to annihilate the diagnosis and secure her release. Unfortunately the boy who could have been related to Mantell's quarryman helper at Whiteman's Green, despite the distance of some thirty miles between Cuckfield and Burwash, was hanged before the matter could be delayed.

In his diary he wrote that 'yesterday morning received the gratifying intelligence that Hannah Russell, the poor woman who was in Horsham jail under sentence of death, was pardoned and set at liberty, in consequence of my communication to the Secretary of State, Mr Peel'. Subsequently Mantell explained the properties of arsenic and his role in the case to the Lewes Philosophical Society, and in 1827 he published 'Observations on the medical evidence necessary to prove the presence of Arsenic in the human body in cases of supposed poisoning by that mineral; illustrated with a case.'

In May 1830 he went with Warren Lee to Burwash and saw Hannah Russell who had remarried. He was greatly interested in her artless and affecting account of her feelings while under sentence of death. She said that the consciousness of her innocence deprived her situation of all its bitterness.

On 31 December 1821, he appeared to have accepted his lot with only a minor grumble. 'Prevented by the usual occupations of the close of the year from continuing my brief memoranda of passing events I terminate another year as I did the last, by anticipating moments of leisure and happiness, which in all probability I am destined never to realised.'

The following December presented a deeper anxiety:

> The past year like its predecessors have fleeted away almost imperceptibly and I am as far from attaining that eminence in my profession to which I aspire, as at the commencement of it, the publication of my work on the Geology of Sussex, though attended with many flattering circumstances has not yet procured me to that introduction to the first circles in this neighbourhood which I had been led to expect it would have done. In fact I perceive so many chances against my surmounting the prejudices which the humble situation of my family naturally excites in the minds of the great that I have serious thoughts of trying my future either at Brighton or London. My little ones however render it necessary that I should pause before I take a step fraught with such importance to them and I am therefore in that anxious state of suspense when nothing can be more unpleasant.

His state of unrequited ambition is further expressed in a poem written at Castle Place in March 1825.

> What if the envi'ous rail, the haughty frown
> My spirit shall pursue its vent'rous way
> And seek in fame's proud temple high renown
> Tho' on my path no kind or cheering ray
> My side encouragingly – I'll reck it not,
> Neglect and scorn to bear too oft have been my lot.
>
> No! whatsoe'er my fate I'll not complain
> If that my humble name enroll'd shall be
> Among the glorious intellectual train
> Whose fame shall live thro' all futurity
> If that the wreath be mine which science turns,
> For those who count her smiles and worship at her stones.

The outburst of 31 December and the carefully penned poetic woes of March 1825 fit uneasily with what he had previously recorded. There is an explanation. His *Geology of Sussex* had been published that September by his brother-in-law, Mr Lupton Relfe. The cost with its numerous plates was

£600 and before printing he had to get subscribers. His Majesty, various professors and the Earls of Egremont and Northampton subscribed but he failed to attract other gentry from Sussex. Consequently his other brother-in-law George Edward Woodhouse, a lawyer, had to put up over £300 towards the cost. Fortunately, the work was in great demand.

> During that summer Mrs Woodhouse, Mr Lyell, Dr Fitton, Sir Richard Phillips etc. have visited me and most of the gentry and nobility have called. In November the Earl of Chichester came and was very pleased with my collection presenting me in December 17th with a fine antique gold ring and Mrs Mantell with some Irish Gold. Mr Curteis, the Tory Member of Parliament also became a frequent visitor.

In autumn 1822 he took on his brother, Joshua (1795-1865), 'whose talents and taste are more suitable to my profession than to his own trade (working in a bookshop) has bound himself as an apprentice to me, for 5 years, at the end of which period I shall send him to the Hospitals.' By the next summer he had sent Joshua to London to study obstetric practice at the Westminster Hospital, and medicine under Dr Armstrong at the Fever Institution. He returned to Lewes to work for his brother leaving again to study surgery at St George's Hospital 1827-8. On 24 April 1828 Joshua qualified as a Licentiate of the Society of Apothecaries.[3] In many ways this was an appropriate qualification; Joshua had previously given public lectures on silver and gases. Licentiates, before the establishment of the degree of LMSSA (Licentiate in Medicine and Surgery of the Society of Apothecaries) in 1815, could legally, by a House of Lords decision, practice physic but could only charge a fee for medications, not for advice or visiting. Popular medications were mercury, antimony, phosphorus, arsenic and iron. Both John Fothergill (1712-80) and William Withering (1741-94) began their medical careers as apothecaries.

Apothecaries could act as general practitioners. To quote John Gregory of Edinburgh, 1772:

> If a surgeon or apothecary has had the education, and acquired the knowledge of a physician, he is a physician to all intents and purposes, whether he had a degree or not, and ought to be treated and respected accordingly.

Dr Thomas Percival, writing on medical ethics in 1893 described the apothecary as 'the physician of the poor in all cases, and of the rich when the distress or danger is not very great'. Yet again in 1810 Jeremiah Jenkins, declared that:

> The apothecary of this country is qualified by education to attend at the bedside of the sick and being in general better acquainted with pharmacy than the physician of English Universities is often the most successful practitioner. The most laborious part of practice falls upon him.

Once qualified, Joshua did not return to Lewes but set up as surgeon in Newick, a village to the north of Lewes. There he founded the Newick Horticultural Society and wrote an essay on floriculture. He remained bright and alert until, in March 1836, he was thrown from his horse irreparably injuring his brain. His brother Thomas auctioned off his scientific and professional effects and in April 1839 he was admitted to Ticehurst Asylum and died in March 1865.

Gideon Mantell's sixty-seven books and papers included several medical reports, nine in *The Lancet*, among which was 'Remarks on partial fracture of the Radius' in 1841. This was one of the earliest descriptions of a greenstick fracture. He had seen six cases, all in children under nine years of age.

On 31 December 1825 he wrote that the close of the year has been so fruitful. On 22 December he records that:

> It was with no small degree of pleasure that I placed my name in the Charter book (of the Royal Society) which contained that of Sir Isaac Newton and so many eminent characters. Mr Babbage and Mr Herschel introduced me.
>
> In the course of the year I have advanced my literary reputation and have been elected an honorary member of the Institute of Paris and a Fellow of the Royal Society, London. My practice has considerably increased, but my strength and health have more than decreased in proportion.

The year 1826 saw more family illness. He was very ill with lumbago in June and in that month Mrs Mantell had a severe illness, showing some recovery in July having been confined to her bed for a month and in September went to Brighton for the benefit of the air and bathing, returning on 5 October. That December, when going to Tilgate, the wheel of his carriage came off and he was obliged to walk many miles, in the dirt and the rain, before coming home in a cart. Then on the 30th, his horse started suddenly and caused a severe rupture on the right side. The hernia remained very painful causing much suffering. Over the next few weeks he became very ill complaining of a tumour in the groin. He possibly had a temporary bowel obstruction. He applied leeches to ease the pain and bought a truss. Eight days later still very ill he used an 'unusual preparation' in the hope of preventing suppuration. On a visit to the city at a later date he went to Wight of London specifically to purchase a better truss. Two months later he was still indisposed but was beginning to recover. However, his 'dear wife' was very ill and all the children suffering from whooping cough. To add to the anxiety and distress he had not heard from his brother, Joshua. Despite the children's illness his wife left for London in the hope that a change of air might improve her health which, for a long time, had been in a delicate state. Though his children recovered, his wife wrote to inform him that she was too ill to return home. The year ended with a heartfelt cry:

> The conclusion of another year terminates by leaving me in the greatest distress possible, not hearing from my beloved Mary today, and unable to leave home to visit her. I am suffering the greatest anxiety and while all my neighbours are either dancing, playing at cards, etc ... and the bells are ringing, I am alone and depressed. Gracious heaven – Relieve me from the pangs I now suffer!

MUSEUM

Relics of a former world
Medals of Creation's birth
Ere history her page unroll'd
Or man was tenant of the earth
(*Sussex Advertiser*, June 1825)

The importance of his fossil collection, the Mantellian Museum, in determining and, in some regards, undermining his career cannot be overestimated. Gideon Mantell collected fossils from that first ammonite found by the river bank. William Constable gave him a mammoth tooth from America. Thomas Woollgar, a Lewes trader with an interest in natural history and medicine, and others encouraged the young child. He continued to collect fossils while in Wiltshire, on his visit to the Isle of Sheppey. He also took his fossils with him to London. His interest was greatly stimulated by his meeting with James Parkinson, whose fossil collection was published in three volumes from 1807 to 1811. Later on James Moore, his mentor and then partner in general practice, and the Woodhouse family, into which he married, added to his collection.

Back in Lewes he made arrangements to pay for fossils found and retrieved from the neighbouring quarries, from Castle Hill at Newhaven, Beachy Head and as far east as Winchelsea. He visited and, through Leney and other quarrymen at Whiteman's Green, collected fossils from the Tilgate forest. These sources, until others tried to outbid him, provided abundant examples. In helping Etheldred Benett with her Wiltshire fossils, 'which contained so many rarities' he gathered yet more fossils, exchanging his surplus ones for more exciting finds. Bituminous wood came from Norwich and through his friend, George Chassereau, he received fossil wood from Jamaica. He assisted John Hawkins (1758-1841) who had moved to Bognor Regis and, though generally more concerned with economic minerals than with fossils, was writing a memoir about the Cornish mines, requiring help in ascertaining the names of the fossils found. Further examples from Cornwall came through Davies Gilbert (originally named Giddy) who was also a Cornish Member of Parliament but came to live near Lewes. By this time he was exchanging finds with James Sowerby who had published his *Mineral Conchology*. Sowerby advised him in October 1816 to publish a catalogue of his wonderful collection of everything found in his neighbourhood. Other senior members of the Geological Society, including the President, G. B. Greenough, came to Lewes to view his museum.

When he combined the two houses in Castle Place, his drawing room with its marble topped table was largely given over to presenting his collection that could be viewed gratis on the first and third Tuesdays of each month from 1.00 until 3.00 p.m. upon previous application by letter. Gideon or a member of the family would be present. In practice, everyone, it seems, dropped by at all times. He welcomed the local gentry and notables, including Lords Egremont and Northampton, listing their names in his diary. Col Birch (later called Bosvile) who had helped Mary Anning by selling her fossils, Elizabeth Cobbold, the poet and naturalist, Sir Astley Cooper, Robert Owen the founder of the Cooperative Society, and eventually Cuvier and all the senior members of the Geological Society visited his exhibits. Bakewell, on 29 September 1829, described 'A Visit to the Mantellian Museum at Lewes' in the *Magazine of Natural History* enthusiastically, praising Mantell's collection of chalk fossils as

> the finest in the kingdom with its splendidly preserved fossil fishes, outstanding ventriculites, remarkable ammonites – but the most interesting were undoubtedly, those from the freshwater Wealden strata below the chalk, including abundant remains of terrestrial plants and large animals.

Meantime his medical practice developed substantially. In the short period when Joshua was in practice with him they saw ninety patients in one day. To do the rounds of the practice he had to travel to the various villages in all weathers and this had an adverse effect on his health, forcing him to take days in bed with chest infections and gastro-enteritis.

The success of his practice meant that he was conferring with distinguished London surgeons and physicians, seeing more and more members of the aristocracy and making daily visits to Glyndebourne and to Brighton. At the same time he did what he could to attend an increasing number of geological events and seize the opportunities to extend his collection of fossils.

By March of 1829 Mantell found it necessary to take steps to minimise the irregular hours involved in his profession, to improve his poor health and to fulfil a desire to have more time for both his family and indulge his avocation for fossil collecting. In his correspondence with Samuel Woodward of Norfolk in the course of 1829 he said that he had very long been an invalid and unfit for anything but had now recovered. And later, that after twenty years of hard work and attacks of illness he hoped to escape some of the drudgery. He agreed a medical partnership of fourteen years with George Rickwood, formerly of Horsham. Mantell would receive two-thirds of the profits for the first seven years and in March 1836 would receive a further £500 from Rickwood who would also pay £900 to initiate the partnership. After seven years the profits would be split evenly between them. This arrangement was renewed with a further partnership between both of them and Andrew Doyle in October 1831.

Part of the money he spent on a more suitable memorial to his parents in St John's churchyard and used the surplus to build a cupola on the roof of Castle Place specifically intended as an extension of his museum, planned to contain uniquely designed cabinets. To this end he compiled a guide of thirty-two pages, published by Lupton Relfe.

> Busily employed in superintending the erection of cabinets in my new room, I received a set of casts of the Ava Fossils from the Council of the Geological Society; also a present from Mr Murchison and Dr Buckland of casts of a femur and phalangeal bones of the Iguanodon.

Bakewell recognised him as the 'proprietor of an outstanding fossil collection'. One wall of this domed room held a pair of antlers of an Irish elk spanning eleven feet. Elsewhere against the wall were five cabinets with an additional cabinet in the centre of the room.

The first case displayed gold, silver and other minerals, recent marine fossils, and highly polished ammonites. Case two was devoted to the strata of Tilgate forest and the Hastings beds with the remains of four enormous reptiles: Iguanodon, Megalosaurus, Crocodile, and Pleisosaurus. The bones of the head, teeth, vertebrae, clavicles, coracoid bones, ribs, chevron bone, femur, tibia, fibula, metatarsals, phalanges, ungual bones and horn. Some of these had been acquired, a few were casts. Other specimens included Pterodactyl, turtles, fishes, freshwater shells and plants. Case three was filled with donated fossils from beyond Sussex: elephants, aurochs, Irish elk, mammoth remains, Ichthyosaurus, a skull of a cave bear, fossil fishes and insects. There were seventy models presented by Cuvier, specimens of natural history from the Paris area. Drawers in case three and all of cases four and five stored British fossils. Under the glass top of case six in the centre of the room he displayed crustacea, his most striking fossil fishes, and some Mosasaurus vertebrae. Its drawers were filled with teeth, shells, and other acquisitions.

The display of priceless relics of matchless beauty clashed with the domestic needs of the house. His wife had long lost any enthusiasm and far from it being a genial family home the atmosphere was cold, scholarly and learned, devoted to the examination, classification and presentation of the fossils. The basement of the house was given over to the domestic staff. The ground floor served as the surgery, and the upper floors shared uneasily between the fossil display and the growing family. To his wife's disgust more and more fossils arrived as gifts, in parcels, packets, and even cartloads. Visiting times were ignored, as he wrote in his diary, 'Worried to death with visitors ... this notoriety is a curse ... my regulations are daily infringed upon to the great annoyance of Mrs Mantell.' Some visitors came to idle away a few hours, others, highly delighted, would stay until the early hours of the morning and yet others would stay overnight. A Dr King arrived with a party of twenty and Count Muster, the Hanoverian Ambassador, came with his lady and family. Mrs Mantell took every opportunity to stay away to the neglect of her children.

With his membership of the various elite societies, his family and his friendships with Lyell and others, he was making regular journeys to London staying two or three nights there. His preferred route was via Brighton using the stagecoach. On St Andrew's Day, for example, which was the anniversary of the Royal Society, he went to Somerset House at eleven o'clock and attended a meeting when the splendid gold medals, given by His Majesty King George IV, were adjudged to Mr Dalton of Manchester for his theory of atomic chemistry and to Mr Ivory for his mathematical papers on spheroids. The Copley medal was given to Mr South for his astronomical labours on the parallaxes of the fixed stars. Sir Humphrey Davy FRS was in the chair and made an eloquent and luminous speech on the merits of the celebrated individuals who were honoured with the medals. He dined with the Fellows of the Royal Society: Dr Fitton, D. Gilbert, Mr Croft, Herschel, Greenough. Mr Peel, then Secretary of State, was present and made a most eloquent address to the meeting. At eight o'clock he went to Mr Lyell's in the Temple for tea, where he met Dr Daubeney of Oxford and Mr Murchison.

Mantell's brooding at the end of 1825 led to a long poem of which the following is but a part:

Thou art fled to the ages past

For what hast thou brought to me
But days of care and sadness
E'en the hopes I cherished in thee
Have left no trace of gladness
And anxious vigils I've kept
In a fruitless search for fame
And have toil'd when the world has slept
To gather a deathless name!

Nonetheless, Mantell took particular satisfaction in the visits of Davies Gilbert and his 'old master and friend, Mr Abernethy and Mrs A. and family'. Mr Abernethy had long been an invalid and was paying a brief visit to Brighton. He looked over the museum and was what Mantell had always found him: kind, intelligent and agreeable. Several senior members of the Geological Society called and came again spending at least three hours on each occasion. They included Murchison with whom he was already familiar, John Phillips, author of the *Geology of Yorkshire*, President of the Geological Society and Keeper of the Ashmolean Museum ('one of the most pleasant, modest scientist I ever met'), and the 2nd Marquess of Northampton (Spencer Joshua Alwyne Compton, 1790-1857) a patron of the arts who became MP for Northampton after the assassination of Spencer Percival (a relative). He was a maverick Tory and minor poet who had lived in Italy and supported Wilberforce in securing the abolition of slavery. A widower, he possessed a large collection of minerals and fossils. Mantell found him a very agreeable and intelligent man. He became President of the Archaeological Institute and President of the Royal Society in 1838. Known as a conciliator, he remained for Mantell one of his closest and most supportive scientific friends.

In 1826 among the gifts which Mantell received in exchange for the fossils of Sussex were: from M. Prevost part of the jaw of a Paleotherium from Montmorency, and from Dr Jaegar of Stuttgard a box of fossils from the Alps. That December he accompanied Murchinson to Scharf, the artist, to inspect the drawing of the fossil fox of Oeningen and went to the Egyptian Hall, in Piccadilly, and saw the Siamese Youths. The lads were about eighteen years of age and united by the xiphoid cartilage of the sternum. The boys were looking healthy and cheerful but did not betray much intelligence, their general appearance was far from agreeable. He returned from London feeling unwell having been driven there seated and holding on to the outside of the carriage in bad weather.

The year ended on a more optimistic note:

> I terminate the year as I began it in my study. I have built a museum and newly arranged and enlarged my collection, published a catalogue of the Museum and the list of Organic Remains of Sussex has just appeared in the Transaction of the Geological Society. I have certainly materially enlarged the circle of my literary reputation and have now foreign correspondents of the first order.

There were other less expert collectors who were keen to show off their collections. William Devonshire Saull, a wine and brandy merchant, had accumulated some 20,000 specimens – about the size of Gideon's – at 15 Aldersgate Street which he opened to the public every Thursday. Mantell exchanged fossils with him in November 1830, befriending and encouraging him in the development of his hobby; the result of which secured Saull's election to the Geological Society in the following year.

December 1829 was momentous in several ways. Buckland read a paper to the Geological Society, 'On the Discovery of Fossil Bones of the Iguanodon in the Iron Sand of the Wealden Formation in the Isle of Wight and in the Isle of Purbeck' with very fulsome acknowledgement of the researches of Mantell and jokingly compared the Saurian's enormous size to that of the small, genteel lizards of our day. There was great excitement and expectation about the proposed visit of Baron Cuvier to England but the brief trip had abruptly to be curtailed because of political upheaval in France. Mantell was determined to see him and carrying a box of fossils took the coach from Brighton to Charing Cross but failed to find him at his hotel. The next morning at 9.00 a.m. hiring a cabriolet, he went with Walter to call on Professor Georges Cuvier. The baron received them both with open arms. Mantell described him as a stout, square-built man with a large face, a noble expanse of forehead, and bright penetrating eyes. After examining and sharing opinions on the fossils they parted, Cuvier assuring him that he would return (which unfortunately never happened). Mantell related that

> I left this distinguished man with great regret. We had never met before, and he was the idol of my scientific idolatry. In truth, the whole time that I was with him I was in a state of feverish excitement which I cannot describe.

He had had similar feeling twenty years earlier when he first met James Parkinson.

One visitor to the museum particularly keen to view the Maidstone Iguanodon was 'Mad' John Martin (1789-1854), 'the celebrated – most justly celebrated – artist, whose wonderful conceptions are the finest productions of modern art', whom Mantell also befriended and persuaded to prepare a frontispiece for *The Wonders of Geology* in 1833.

Martin is said to single-handedly invented, mastered and exhausted an entire genre of painting, the Apocalyptic Sublime. No artist exerted more popular and powerful paintings of Apocalyptic visions which were hung in the famous houses, Stowe, Deepdene, Howick and Buckingham Palace. His pictures were enormous, grand-eloquent, obsessed with spatial infinity, and satanic heroism, influencing the Romantic imagination of his day. Titles such as The Deluge, The Expulsion of Adam and Eve, the Seventh Plague of Egypt, The Creation, and Pandemonium were presented theatrically and reproduced as etchings, engravings, book illustrations and mezzotints. The atmosphere of The Deluge was reflected in a poem by Bernard Barton (1828):

> This Awful Vision haunts me still
> In thoughts by day, in dreams by night;
> So well hath Art's creative skill
> There shown its fearless might.

He also used his skills as a draughtsman to improve public amenities, including the Thames embankment and a bridge across the Menai Straits.

Born in Northumberland, he was one of four brothers, all either mad or eccentric. Jonathan climbed into York Minster and spent four hours getting a fire going with prayer books and furnishings and ended his days in Bethlem Hospital. Richard published a book of poems, and William became an inventor, Christian philosopher and pamphleteer. Many of Martin's scenes have been ascribed to the effects of opiates but more probably he was obsessed with the effects of magic lanterns, coloured glass and dioramic displays. In producing the frontispieces for Gideon Mantell he was able to examine the museum habitat group by William Bullock and a lithograph, 'Duria Antiquior' by Henry de la Beche, to show a country clothed in tropical vegetation, consisting of arborescent ferns, zamias, and coniferous trees as in the fossil remains of Tilgate forest. There were Jurassic oolites forming cliffs in a landscape traversed by a river with a distant sea. In the foreground were an Iguanodon, Megolsaurus, Crocodile, Hylaeosaurus and ammonites; and among the trees a Pterodactyl. The scene thus closely reflected Mantell's depiction of the possible landscape.

In general Lyell's opinions and advice were gladly accepted by Mantell. Lyell realised that Mantell was trying to establish a geological career while managing a busy medical practice and wrote to him urging him to cut out botany and geology to concentrate on British fossil reptiles and fish.

> I have sworn to myself that you shall show them ere many years who and what you are and put to blush the jealous unwillingness which most metropolitan monopolists in Science, both in France and here, exhibit towards all such as happen not to breathe their exclusive atmosphere. But you must concentrate ... cut out botany ... give up all ideas of a popular book on geology ... But from this moment resolve to bring out a general work on the subject of 'British Fossil Reptiles and Fish'... Clift has no time; Buckland is divided amongst a hundred things and is no anatomist.. You must say nothing for a while. By this you may render yourself truly great.. the field is yours but might not remain open many years. It is worthy of your ambition and the only one which in an equally short time you could make your own in England and forever.

Following this advice Mantell began a remarkable paper in November 1829 on 'The Age of Reptiles'. These enormous quadrupeds were the Lords of Creation before the advent of the human race.

> There was once a time, as geologists say
> That reptiles alone o'er our planet had sway
> And the 'Lords of Creation' were all creeping things
> Some crawling on earth and some soaring with wings!
> These monsters so greatly polluted the air
> That nothing but reptiles inhabited there
> Till a grand revolution destroyed the whole race
> And refined earth and air, and then men took their place
> So Cuvier hath taught, and by Mantell more lately
> It seems (if he do not exaggerate greatly)
> That their size was appalling and far surpassed all
> The monsters that lived since the time of the fall.

The Rev. William Kirby protested, in words beloved of modern-day Creationists, that:

> Who can think that a Being of unbounded power, wisdom and goodness should create a world merely for the habitation of a race of monsters, without a single rational being in it to glorify and serve him?
>
> (*On the Power, Wisdom and Goodness of God as Manifested in the Creation of Animals, and their History, Habits and Instincts*, 1835)

In June 1830 he was very pleased to receive a visit from his distinguished relatives Sir Thomas Mantell and Lady Mantell of Dover. 'Sir Thomas who has attained the age of 81 is an extraordinary man for his years and

Lady M is as delightful and active a woman as it is possible to conceive a female can be after being married above 50 years'. Sir Thomas, (1751-1831), like his father trained as a surgeon, but after holding various posts in the port of Dover was knighted in 1811. He was an antiquarian who investigated tumuli and published an account of the Cinque Port Meetings. The following 19 April Gideon Mantell returned his visit taking the Dover coach and putting up at the London Hotel notifying Sir Thomas Mantell of his arrival. The old gentleman was delighted to see him. There had been an earthquake that March. As he commented in his journal, 'it is curious that since the Conquest the earthquakes in England have been most frequent and severely felt on the Kentish coast'. Had the earthquake loosened the rocks of the cliffs? He was emboldened to rise the next morning at 4.00 a.m. and, with his hammer in his hand, strolled along the beach to Shakespeare's Cliff. He later breakfasted with Sir Thomas, looked over his library and antiquities and visited Dover castle and fortifications. On the 22nd at the same early hour and in company with Sir Thomas and Mr Pearce, his secretary, they went in a chaise to Folkestone, alighted at Eastware bay and walked along under the cliff collecting fossils of the Galt and Shanklin sand on the way. They breakfasted at an inn and had a delightful ride back to Dover obtaining a good view of the coast of France from a hill near Dover.

Then at noon Sir Thomas and Lady Mantell accompanied him to Canterbury in their carriage arriving in the city soon after 2.00 p.m. Lady Mantell went with him to the museum where they were gratified by the opportunity to inspect many beautiful and rare fossils. They then went to the Cathedral. Mantell was struck by its grandeur. They found that the Neville chapel in which the Mantells had been buried had been destroyed and the monument of Ann Mantell removed to the Dean's Chapel and placed in a very obscure situation. He left Canterbury at midnight and arrived in London at six the next morning. On 23 April he met up with Mrs Mantell and her mother in Kentish town, called on Relfe, his in-law, and Somerset House and left London with his wife at 3.00 p.m. reaching Castle Place safely at 9.00 p.m.

With the greater freedom of movement with others sustaining the practice he went to London on 7 June by night coach from Brighton, breakfasted with Dr Hodgkin[1], called on Mr Bell and Relfe then proceeded to see Sir Astley Cooper and had a long gossip with him. He then drove to Langham Place and had an interview with Sir James and Lady Langham. He then went to Mr Murchison and took luncheon. Afterwards, he accompanied Relfe to the zoological gardens and then to Sowerby's in Regent Street and Leadbetters in Brewer Street where they saw some splendid birds, many of them unique. He dined at a restaurant in Leicester Square and finally took coffee at York Hotel. On the 9th he went to Bromley, then to Somerset house and called on Mr Lyell spending the evening with him. On the 10th after breakfast with Mr Lyell, and his cousin Dr George Mantell, he returned to Lewes.

Once back in Lewes he went to Brighton nearly every day with visits to Lansing Down to see the remains of a Roman Temple, also making further geological excursions to Newhaven collecting shells, to Steyning for fossils, Bramber where he received an unidentified animal bone and to Cross in Hand to examine Paludrina – on each occasion returning between nine and eleven in the evening. He frequently went to the theatre, in July with Mrs Mantell and saw Kean in Brighton, acting the role of Sir Giles Oversuch, before retiring. Mantell gossiped after the show with Kean. He met his cousin, George Mantell, once again, when with his three sons, he came on a visit from Farringdon in Berkshire. In the same month he received a splendid model of the head of the Mososaurus from the Musée de Paris as a present from Baron Cuvier.

King George IV died on 1 July 1830. The accession of William IV was greeted with approval as he was believed to support the Whigs, though later he reneged on the Reform Bill. With a break in tradition the new king and queen took up residence in the Pavilion at Brighton arriving in September. As the *Sussex Gazette* reported, 60,000 people, an almost countless throng, gathered to welcome their majesties on that joyous occasion. Furthermore they were to pay a visit to the county town of Lewes. The accession brought on Mantell a sensational, almost melodramatic, change in attitude. From his journal, on 6 September, in advance of the king's visit to Lewes he explained:

> This evening wrote addresses to the King and Queen, resolutions, etc. for the town meeting. I am indeed jack of all trades – more fool I – for I get neither profit, credit nor thanks, still there is a pleasure in moving the public mind and guiding it unseen.

On 22 September William IV and Queen Adelaide paid a visit to Lewes. The royal carriage arrived from Brighton at 12.30 p.m. entering Lewes by the Western Road. The streets were gay with flags and streamers and the houses all decorated with greenery and flowers. Everybody was there to welcome the first official visit of the sovereign to Lewes since the Middle Ages. A grand banquet was held at The Friars and a fine armorial dinner served on Worchester china made to commemorate the event.

In the *Town Book of Lewes* there is no mention of Gideon Mantell. The king and queen were greeted by the MPs, Sir John Shelley and Thomas Read Kemp, with their chief officers and headboroughs. Loyal Addresses were given and the king replied. After the meal the king went to County Hall and the queen and princes saw the Castle before returning to Brighton. From Mantell's journal we have a very different account:

> The King and Queen visited Lewes. Their Majesties had signified their intention of seeing my Museum and we therefore had everything prepared for their reception, but the various ceremonies occupied so much of the time that it was too late before they completed them and a message was sent to me that their Majesties would honour me with another opportunity. I was presented to His Majesty at The Friars by Sir John Shelley in the presence of the nobility, etc. On my offering the 'History of Lewes' to his Majesty I knelt on one knee and said, "Sire, Your Majesty's gracious condescension, in deigning on a further occasion to accept my Geological Work induces me to hope that Your Majesty will permit me to lay at your feet, the History of this my native town." His Majesty almost interrupted me with; "Certainly, certainly, much obliged, much obliged" and turning to the Lord Howe in waiting said in his usual hasty manner, "take them, take them". Afterwards he met Sir John Shelley who told him Her Majesty intended to come and see the Museum in the Spring.

His *History of Lewes* was based on six letters he had written in 1813 describing the Battle of Lewes, the castle and priory, the Friars at the Hospital of St John, St Nicholas church, the pigeon house at Lewes Priory and the other churches of Lewes, Southerham, Cliffe and Malling. There were drawings of Lewes from Old Hithe, Offham road, and the Coombe along with the ruins of the Priory, the east view of the Western Keep of the castle, and the monument of Magnus on the south wall of St John's church. He also had his relation Lupton Relfe publish *A Narrative of the Visit of their Gracious Majesties William IV and Queen Adelaide to the Ancient Borough of Lewes on 22nd October 1830*. This work he dedicated to Sir Thomas Mantell of Dover.

Though the king did not subsequently visit the museum in Lewes, on 8 February the two young princes, George of Cumberland and George of Cambridge, with their tutors, the Reverends Jelf and Wood, came over from the palace at Brighton to see the collection. Prince George, aged 13, a fine boy for his age and very like the late King George IV and Prince George of Cambridge two or three years younger were both shrewd, intelligent lads and very pleasing and affable in their manners and conversation. They were both highly delighted with the various objects in the museum and comprehended what was said to them about geology. They stayed for nearly two hours.

To commemorate the visit to Lewes by the king, Mantell wrote a narrative of the visit and later sent a box to Sir Herbert Taylor containing elegant copies of the book for her Majesty, the two young princes and Sir Herbert. In response Mantell obtained a letter back from Sir Herbert informing of the presentation of the book to the king who was most gratified by it. Furthermore he sat – or rather stood – for the artist Archer to paint his portrait to commemorate the king's visit to Lewes. He intended that the picture would be for 'my dear boy Walter when he is of age – or should his life not be spared – for Reginald, the rioter'.

After a brief visit to London he returned to find Lewes all bustle on account of the approaching election. The king had dismissed parliament the previous night. The narrative of the Royal Visit to Lewes had been published but so entirely was the public attention engrossed by the Reform Bill and the election that scarcely a single copy had been sold.

BRIGHTON

Thoughts of relocating from Lewes to Brighton showed themselves in Mantell's diary entry of December 1822. He still had a wide field of medical practice but after twenty years of hard work and attacks of illness he hoped to escape a good deal of the drudgery of a country practice and the almost incessant loss of rest which his obstetric practice involved. He hoped to secure a more selective and lucrative clientele. Spokes suggests that Mrs Mantell's illnesses and domestic trials influenced her desire to seek a move from Lewes.

Brighton had become the fashionable resort, gaining a reputation for the convalescence and health-giving properties of sea water. The king's brother, the Duke of Cumberland first visited Brighton in 1771 finding benefit to the swelling of glands in his throat from bathing in sea water. Others such as Dr Johnson also stayed in Brighton to test the claims of its medicinal wealth. The Prince Regent made it his playground, and it became the more or less permanent residence of the heir to the throne. Here, as with Beau Nash in Bath, the Prince assumed the role of the arbiter of fashion.

He built the Royal Pavilion which underwent several metamorphoses. As Prince of Wales he leased a small house and in 1787 had Henry Holland reconstruct it as a classical villa with a central rotunda and dome. Then in 1812 Prince George, now Regent, called on his Surveyor-General, the architect John Nash, to add a great onion-shaped dome, tent-like roofs, numerous pinnacles and small minarets to the pavilion. The Chinese style interior was decorated with moulded ceilings, be-dragoned chandeliers, and elegant furnishings. Not everyone was impressed: William Cobbett declared it a combination of a square box, a large Norfolk turnip and four onions. The stone for the building came from the quarries at Whiteman's Green near Cuckfield. Cuckfield, to the north of Brighton, was also the staging post to London enabling faster and safer travel from Brighton to the metropolis than cross country from Lewes. Mantell regularly travelled to Brighton to catch the coach to London.

Brighton had become a cosmopolitan town. Despite the original summer attraction of sea-bathing, with the succession of William IV the official Brighton season shifted from summer and lasted from October to March. The regular appearance of the court, and the excellence of communications from the capital, increased its popularity and added to the lively animation of the scene. By 1815 coaches took only five hours to accomplish the journey each way and in 1818 fifty-two coaches ran daily to London at a cost of six shillings. In the years between 1801 and 1831 the town expanded from 1,282 houses to 7,770, and in 1841, with the opening of the railway, there was an even greater influx of people. To quote Sydney Smith, 'I think all rich and rational people living in London should take small doses of Brighton from time to time.'

In fashionable Brighton – according to Osbert Sitwell and Mary Barton:

> every upstart of fortune, harnessed in the trappings of the mode, presents himself to Brighton ... knowing no other criterion of greatness but the ostentation of wealth, they discharge their affluence without taste or conduct, through every channel of the most absurd extravagance, because here, without any further qualification, they can mingle with the princes and nobles of the land.

The scene at Brighton had reached a high point of fantasy. The society which gathered round the Prince worshipped no other god save Pleasure.

One of the attractions of Brighton for the hoi polloi was its Chain Pier, the first edifice of its kind in England, nearly a quarter of a mile in length, constructed by Captain Brown, RN in 1820. The platform was supported by chains anchored into the cliff at one end and into rock in the sea at the other, and suspended from towers. Modelled on the gateways at Karnak, it stood 13 feet above the high water mark and contained several small shops, fancy ware sellers and a profile artist. Below the pier head was a gallery where in a high sea the breaking of the waves may be seen and heard to great advantage; merriment was often excited by hasty runs and jumps to avoid showers of spray. Coal and merchandise could be landed and people no longer had to be taken ashore from boats in punts. The steam packet to Dieppe arrived and departed from its far end, bands played and fireworks illuminated the night sky.

Mantell's interest in the Chain Pier began during its construction. On 6 September 1823 he walked, with Mrs Mantell, to the end of the pier one week after the flooring had been completed and two months before the official opening. He visited it again from Lewes on 23 November 1824:

> A severe hurricane coincided with the Spring tide. The sea was raging with great violence. So soon as the water had retired to allow walking on the esplanade (we) went to the pier which was very much damaged by the waves, and the platform destroyed, so as to render access to the pier head difficult and dangerous. However we ventured to the furthest end and although every now and then the sea dashed over us and completely drenched us; but the awful grandeur of the scene more than compensated for the inconvenience of the situation.

The pier survived the ordeal but was severely tried by a storm and found fully competent to endure any such convulsions. The waves had been mountains high, often obscuring the pier from sight, breaking the wooden railings and occasionally raising the wooden platform several feet between he towers, but from its elasticity it speedily recovered its proper place and no part of the chains or piles was broken.

For Mantell, the one royal visit to Lewes from Brighton in October 1830 helped to impress on him his town's inferiority to Brighton. The visit was so brief that the king did not have time to visit Mantell's museum. Moving his fossil collection to Brighton would not only render it more accessible to interested visitors, both geologists and gentry, but further his scientific reputation. He had worked very hard to set up his medical practice, which remained a country practice where 'like a torch I consume myself'. He sought the advice of Bakewell and others whom he trusted in contemplating his move. There was the opportunity in Brighton, with its wealthy aristocrats as his clientele, to build up a practice which would be both less demanding and more prosperous. Furthermore he hoped to acquire an honorary MD to establish himself as *the* practitioner in Brighton. 'In confidence,' he wrote to Silliman on 3 October:

> I have almost resolved to move to Brighton, where there is a larger field for my professional engagements, and the chance to turn my success to some account. Another week has passed away, alas, how uselessly! Shall I leave this dull place and venture into the vortex of fashion and dissipation at Brighton or shall I not? Prudence when a man has turned forty with four children and but a very modest fortune removal is a subject of no little anxiety, particularly to a mind like mine but *Ambition* says go and prosper! What shall I do?

The rise of Brighton was not the only reason for the decline of Lewes. The fashionable set had moved to the continent as well as to Brighton, leaving, as the *Brighton Gazette* of 1830 recorded, too many (Lewes) townsfolk engaged in commerce. The improvement of Shoreham as a port had been at the expense of Newhaven. The Lewes practice had been the sole source of Mantell's income for almost twenty years and could fall commensurate with the decline of Lewes. As he told Professor Silliman, 'when I reflect on the many hundreds of families whom, even in my comparatively short life, I have seen reduced from affluence to poverty, I shudder with horror lest such a fate be mine.' Even his own wife's brother-in-law, Lupton Relfe the publisher, had become bankrupt in the early 1830s, and his plight and that of his wife was a continuing concern, one of misery and wretchedness.

The lower classes in the countryside south of the Thames suffered from the harshest of predicaments and this was certainly true of the neighbourhood of Lewes. The years from 1815 to 1820 after the Napoleonic wars, were among the worst in the nineteenth century as desperate workers faced unusually harsh winters, declining wages, massive unemployment and an increasingly oppressive government. The harvests from 1792 to 1813 were exceptionally bad. The Speenhamland Act supplemented the wages of the very poor out of the parish rates. Every poor person was entitled to receive a scale of payments from the parish to make up the deficiency of his wage to 3*s* a week for himself, and for every other member of the family 1/6*d* a week, when a loaf cost 1*s*. As the price of a loaf rose higher the dole was to rise with it.

The Speenhamland Act meant that on the large farms and estates labourers could be paid a reduced wage, often in the form of inferior corn or beer. Small independent farmers, not necessarily employing labour, were crushed by the enactment of the parish rates. These factors were further exacerbated by a rise in the birth rate and the plentiful supply of labour. In 1815 the Corn Laws prohibited the import of corn until the price of wheat stood above 80*s* a quarter. It was a rural, agriculture depression with a lesser effect in industrial areas. Many of the poor survived eking a living by poaching, despite severe penalties. In the era after Waterloo rick-burning was a common form of vengeance commanding the secret sympathy of the whole village. With the ending of the war soldiers were demobilised and barracks, as at Ringmer, closed down. Starving field workers became restless and in the winter of 1830 rioted to demand a wage of half a crown a day. In the countryside south of the Thames judges took terrible revenge. Three rioters were hanged and 420 deported as convicts to Australia. As reported in the *Town Book of Lewes*:

> In the course of the autumn of 1830 an incendiary spirit originating in the want of employment for, and the low rates of wages given to agricultural labourers first showed itself in the county of Kent from whence it spread to this county ... Numerous barns and other farm buildings as well as stacks of corn and hay were destroyed by fir and large bodies

> of agricultural labourers tumultuously assembled together in different parts to demand an increase in wages in consequence thereof the whole county was thrown into an unprecedental state of alarm and confusion from which this Town did not altogether escape.

Mantell describes the events in Lewes. He was confined to bed with illness all day, but soon after twelve in the night, was woken from a deep sleep, produced by a strong opiate which his complaint rendered necessary, by loud cries of fire and the ringing of Bellman's bell. He jumped out of bed and on looking out of the window perceived a glare of fire in the direction of Southover and learnt that Mr Moore's barn on the Priory farm was in flames. Mantell's sympathy with the lot of the agricultural labourer is clearly stated in the entry of 31 December 1830: 'It is all bad, our peasantry are in a state of positive ignorance and slavery, almost starving, without the knowledge necessary to enable them to attempt obtaining redress without violating laws'.

Riots continued, soon afterwards barns were burnt at Beswick, culminating in the Swing Riots of 1835 affecting Kent and Sussex. Close to Lewes rioters assembled at Ringmer and troops were about to be sent to disperse them. However, a local landlord acquiesced to the labourers' demands and so saved the day.

If his sympathies were with the lower classes, his medical practice to an increasing extent involved the gentry. One particular patient, the Hon. Miss Langham of Glyndebourne, he saw nearly every day from 1830-31. Her disease, though not described, appeared intractable from the start and various second opinions were sought. On 11 November after a visit with his wife to Brighton he went in the evening to Glyndebourne to meet Sir Henry Halford, President of the Royal College of Physicians for over twenty years and Physician to the King. He had come expressly to visit Miss Langham and complimented Mantell very much 'I know not what' (admitting that he had no clearer diagnosis) and on his treatment of Miss Langham and the lucid description of her case he had sent him. After dinner with Sir John and Lady Langham, Sir Henry entered into gossip on his Majesty's etc. and his other royal patients. Sir Henry (according to Mantell) is about the middle size, and must be near fifty-five years of age, although looking much younger and has an exceedingly pale and elastic appearance[1]. George IV would never allow him to be out of his sight one moment for the last three months of his life, except six hours to sleep and one hour in the morning. Again on 15 January, Sir Henry called at Castle Place at one o'clock in his carriage and four and went to the museum before being accompanied by Mantell to Glyndebourne. On the trip he read a paper he intended to deliver on the influence of the body on the mind. On

23 April 1831, Mantell rose at 4.00 a.m. and drove to Glyndebourne to breakfast with Sir Henry. The prognosis on Miss Langham was considered poor and he had a long gossip with Sir Henry about cholera.

However, Miss Langham improved that August and took an airing in her carriage. But in October 1831 Mantell drove to Glyndebourne and attended Miss Langham in their carriage to Brighton:

> One of the most anxious moments that fell to his lot. The poor girl was very much affected by the action of the carriage that it was with difficulty we could keep her alive although we stopped every hundred yards. Arriving at 31 Marine Drive he took his charge in his arms and carried her upstairs and laid her on her sofa and never did I feel more relieved from anxiety than then.

Another opinion was sought from Mr Brodie (Sir Benjamin C. Brodie, President of the Royal College Surgeons, and a member of the Royal Society), described by Mantell as a short, little man, of a pallid and unhealthy appearance, with a penetrating eye, but insignificant appearance, affable and polite. He had a good reputation but could offer no alternative means of treatment when he saw her in December. Mantell's daily visits to Brighton ended in May 1832 with the death of Miss Langham. He happened to arrive one day just as she was seized with convulsion and remained with her all night and the next day until she expired in his arms.

On 30 December 1831 he had more success in answering a call from General Trevor at Glynde. Captain Codrington, son of Admiral Sir Edward Codrington, the hero of Navarre, had received a shot in the eyelid whist rabbit shooting. It was necessary for Mantell to cut down and extract the buckshot, which he accomplished very easily.

In June 1832 he learnt that he had been elected as a corresponding member of the Academy of Natural Science of Philadelphia, and in the same month took full advantage of his partners in general practice to attend scientific meetings and geological sites. Thus on 21 June he left for Oxford and arrived in time for breakfast with Dr Buckland. At eleven o'clock he proceeded with Lord Northampton in his carriage to the London entrance of Oxford and on the bridge found a large concourse of persons, some in carriages, on horseback and on foot to meet Dr Buckland and accompany him in his geological ride to Shotover Hill. The morning was beautiful and the scene most animated. Dr Buckland came and jocularly named Mantell as surgeon to the expedition. All Oxford was in attendance and once out of the city the professor began his harangue, pointing out the diluvian deposits in the valley of the river in the most excellent style imaginable. They followed him from one station to another and joined up with many acquaintances. Next they made a visit to quarries and returned to Corpus Christi College in heavy rain just in time to avoid the wet and dress for a very sociable dinner. In the evening he attended the Geological Section and afterwards accompanied Dr Fitton to Merton College to read through Babbage's challenging book on machinery, which was just out. He parted with Dr Fitton at two in the morning.

Over the next four days he attended the Geological Society meeting. There were lectures in the Theatre of the University, visits to the Ashmolean Museum and purchases from Sowerby's of a terebratila (a stinging or boring fossil) and a trigonia (a fossil found in Gloucestershire limestone). His own talk was on the Wealden formation and the saurians; expounding his proofs of the freshwater origins of the strata and illustrating the structure of the Iguanodon from its teeth, bones and claws. Among his exhibits were drawings of a Hippurite (an extinct bivalve mollusc present in chalk) found in the Lewes chalk – the first found in England, though he commented that it was an engraving of a hippurite in an article on petrifications in the *Encyclopaedia Britannica* when he was a boy of 13 that first drew his attention to fossils. On the final evening Dr Buckland gave a lecture on Megatherium.

On the Sunday morning he rose early and called at Christchurch to take leave of Lord Northampton, Mr Whewell and Dr and Mrs Buckland; thanking Buckland for his liberality and generous attention, for which Mantell felt he could never be sufficiently grateful. He left to accompany his cousin to a family funeral at Farringdon and thence to Bristol finding to his great delight that Mr Conybeare and his son were fellow passengers. Stopping on the way at Swindon, he saw once again the long walk – the scene of his boyish rambles and the field where he used to play cricket. Finally passing through the richness of the Valley of the White Horse and its coral rag he arrived at Bristol. Here Stutchby was on hand to show him the museum laying out its geological treasures. He went on to Clifton, Lyme Regis, Knowle Hill and Bath. 'Bath, beautiful, delightful Bath'. At Pensfold he was in a carriage with a Mr Jesty whose father had practised vaccination as early as Dr Jenner, and was himself the first person to be vaccinated intentionally. His father was sent to London by opponents of Dr Jenner and remained there a considerable time until it was obvious that parliament would not grant any money to any claimant but Jenner. Mantell found Mr Jesty a shrewd and intelligent farmer and was amused greatly by his droll remarks.

Towards the end of June he made a further excursion to the West Country visiting the limestone of the Mendips, Glastonbury and Shepton Mallet and staying at Castle Carey, a very pretty village. There was a grand rustic fete in honour of the passing of the Reform Bill. The bells were ringing and in a field at the roadside were booths and a band of music. He proceeded to Cadbury, observing the layers of white oolite, Yeovil and Taunton, where the valleys contained red marl, chalk flint and boulders of chert. Thence on to Lyme Regis to visit Mary Anning, Portland, Chiswell Bay and the petrified forest with stems of trees thirty feet long and fossil cyclads. From Weymouth he took the Southampton coach to Dorchester, Blandford and Christchurch arriving at five in the afternoon. Eventually he returned to Lewes having 'had the longest holiday he had ever had.'

Later that year he made various excursions over several days. On 3 December he left for London with his large Iguanodon bones and his celebrated new fossil. Mr Warren Lee had done a fine painting of the hind limbs of the Iguanodon based on a careful calculation of the comparative bulk of the muscle and its integuments. On the 5th, in preparation for the evening's lecture, Dr Buckland joined him and he read the principal part of his paper and pointed out the characters of his fossils. At 8.00 p.m. to a very full meeting with all his friends present including, Bakewell, Chassereau and Weekes, he was directed to curtail his talk by a third. However, all passed off very well and he concluded by showing the painting which very much gratified the greater part of the audience. Then, to his astonishment, Dr Buckland commented on part of his statement and assumed that he (Dr Buckland) had suggested that the spines belonged to the back of the animal. Mantell comments that this was wholly unlike the candid and liberal manner in which he had always acted to him. Mr Clift, Owen and all the comparative anatomists agreed with Mantell.

Over the next few days he attended a very dull meeting of the Royal Society and the Linnaean Society, arriving back in Lewes just after the Lewes election was over. On return to Lewes, every minute not engaged in his profession was devoted to his book, which he completed in less than three weeks. He had written (much of which was compiled from former volumes) 280 pages with 130 subjects drawn for woodcuts. Professor Silliman sent him a description of his Lewes museum which appeared in the *American Journal of Science* for October. January 1833 he sent the book to Mr Murray to be published. On returning that evening, Sir Astley Cooper arrived at his house to see his patient, a Mrs Newton of Southover who was suffering from gangrene of the foot but was showing some improvement. He accompanied Sir Astley to Mrs Newton's. He approved the plan of treatment and after seeing the museum returned to London. Later that month Mr Murray declined to publish the book, Relfe could not; and Bakewell suggested sending it to Longmans, his own publisher. In February he received a letter from Sir Herbert Taylor stating his Majesty's desire that the new work be dedicated to his Majesty. He had already informed Sir Henry Holford of his intention to inscribe it to Sir Henry.

On 13 March on a very cold morning he visited four or five patients before taking the nine o'clock coach to London where, with Mr Bakewell, he visited Longmans. The next day, he saw Mr Hawkins and his wonderful specimen of the Ichthyoplatyodon; afterwards going with Mr Lyell to the zoological museum. On the 16th he went with Saull to a sale of Egyptian antiquities, and thence to a conversazione at Kensington palace presided over by the Duke of Sussex (1773-1843) the sixth son of George III, famous for his hospitality and President of the Royal Society from 1830-38. Mantell was presented to His Royal Highness who received him most graciously. Four hundred persons were present from among the most distinguished in science, literature and rank. He was highly gratified at not only the flattering reception but with the opportunity to meet and see some of the most celebrated characters of the age. Michael Faraday was showing an experiment involving magnetism, when a rather decrepit man approached in court dress and ruffles, talking in French and took part in the experiment. Later on Mantell found himself talking to the same old man, 'a tall man, plainly dressed, of plain features, yet with a sly expression of deep penetration and reflection, sitting on a sofa with Mr Brodie, the Surgeon, and Lord Brougham, the Lord Chancellor', who turned out to be Prince Talleyrand. Mantell admired a splendid bust of Napoleon and discussed geology and astronomy with Sir. J. Herschel before sitting to dine with Lyell, Clift, Horner, Konig, Greenough and Lambert. He retuned to Lewes the next day after breakfasting and gossiping with Dr Buckland.

On 8 April he went to London by coach with Walter and called on Chassereau. He then visited Sir James Langham at Langham Place whom he found in a dying state and waited a long time for Sir Henry Holford and Mr Brodie. Later he went to the Coliseum where he joined Walter and Mr Chassereau. He called again at Langham Place the next day. On the 15th he received a letter from Mr W. S. Langham to say that his father had died and he then communicated the death to his son, Sir J. B. Langham at Glyndebourne.

In May Mantell made further trips to Pulborough to meet up with Mr Saull and called on Peter John Martin (1786-1860), surgeon and geologist. After seeing the Pulborough quarries he had breakfast the next day at Petworth. Later he spent three hours looking at the collection of statues and paintings of the Earl of Egremont, proceeding to the Roman villa at Bignor and staying overnight at Arundel. He returned a day later via Worthing. In June he went to Cambridge having sent the femur of the Iguanodon ahead by van. Lady Mantell stayed with Mrs Mantell at Castle Place. He visited his sister Kezia in Bromley, then went with Mr Davies Gilbert to the Philosophical Society and on to a grand dinner at Trinity College Hall. In the course of several days of varied activities he unpacked the thigh bone of the Iguanodon to show the Geological Society, breakfasted with Dr Buckland and Lord Northampton and then departed in his Lordship's carriage to Castle Ashby visiting quarries and staying for three days 'of uninterrupted happiness' at Castle Ashby, where Drs Buckland and Murchison were accompanied by their wives, so that, as Mantell recorded: 'We had a most charming party and I very much enjoyed myself.' His journey on to Oxford was via Northampton, Towcester and Stowe. The Duke of Buckingham from his palace at Stowe had purchased the first Plesiosaurus. Mantell was suitably impressed. Of Stowe, he wrote, 'never have I seen so charming a spot, nor so splendid a palace. The scene was indeed truly classical: temples, fountains, lakes, statues; what a fairly scene!'

*

Among those with whom Mantell had discussed his removal to Brighton was the third Earl of Egremont, (1751-1837), Lord Lieutenant of Sussex. Though a patron of the arts and civic betterment, Egremont surprisingly had decided not to subscribe to Archer's painting of 'the loyal and liberal inhabitants of Lewes'. His thoroughbred stable completed each year in the race weeks at Lewes and Brighton. As early as 1821 he had been one of the most prestigious subscribers to Mantell's *Fossils of the South Downs*. He had called at Castle Place in December 1825 to congratulate Mantell on his election to the Royal Society. They reciprocated visits the following May 1833 when Egremont inspected the Castle Place Museum and Mantell spent three hours at Petworth, the Earl's magnificent estate in West Sussex with his outstanding collection of paintings by Leslie, Turner, Constable, Carew, and others, and sculptures and drawings by Flaxman. His artistic patronage, horse-breeding and civic generosity displayed an extrovert personality. The elderly lord was not only very rich and powerful, but had led an unusual life. He had six children by his mistress, Miss Iliffe, eventually marrying her in July 1801. A legitimate daughter died in infancy but in May 1803 they separated with a settlement. As John Wyndham, his descendent, wrote:

> What happened, I suppose, was that Miss Iliffe, when she came downstairs as a wife after being upstairs as a mistress, started bossing the servants about and interfering with Egremont's arrangements and disturbing his comfort. He needed a mistress but he could not manage a wife!

Having realised that the move to Brighton was a real possibility Mantell informed Egremont, who responded the next day with an invitation and a

The 3rd Earl of Egremont (1751-1837) from a painting of 1828 by William Derby engraved by H. Robinson and published by Philips at Petworth.

No. 20 The Old Steine, 1833-39. Mantell's residence and museum in Brighton.

gift of game. When Mantell went to Brighton on 5 October to thank him for his gift and expectantly waited upon his Lordship, now an octogenarian, Egremont spoke to him approvingly on the subject of his relocation. Magnificently, in his characteristically blunt, unceremonious manner, Egremont offered Mantell *one thousand pounds* to assist the removal. In making what was to be a fateful decision, Mantell conferred with Lyell and other close friends, and decided to accept the Earl's extraordinary offer. That November he concluded an agreement with a Mr Budd for the lease of No 20, The Steine and worked on the numerous details involved in the bustle and hurly-burly of his removal to Brighton.

By 1 December he had re-established his collection in the new house and substantially re-arranged the drawing room. On the 21st his family and all his servants were installed in their new abode. In the same diary entry he bids farewell forever to Castle Place and writes 'still beyond measure tormented by my late partners who have behaved in the most shameful manner towards me'.[2]

Charles Lyell was confident that Mantell's move to Brighton would meet with success. Between December 1833 and January 1834 Mantell had administered to Lyell's mother and sisters in Hastings in a medical capacity. Lyell told his friend, Dr Fleming of Aberdeen, that:

> Mantell has made a bold professional stroke in removing there, which you will be glad to hear is likely to succeed, in spite of the misgivings of many of his friends, who had not the confidence which I always had in his genius. He is, in fact, a man of great medical skill, and tact so great, as to triumph over the drawback of his having so fine a museum and so much fame in certain branches of geology.

Old Steine formed a triangle around ornamental trees and gardens, lined about with rows of elegant double-fronted houses. The thoroughfare was overlooked by the Royal Pavilion above and the Old Ship Inn and Palace Hotel towards the seafront. No. 20 The Steine, was an imposingly tall but narrow structure of several stories with a splendidly swollen bow-fronted façade, decorated within by elegant carvings and marbled fireplaces. He had deliberately selected a house larger than his needs and more expensive than his means at a cost of more than £350 per year in rent and taxes. 'Where better', he thought, 'to place his Museum: the first to show the three known giant land reptilians'. For over a month from May 1834, he sought to assemble the Maidstone Iguanodon which had arrived in a fragmentary and shapeless state. He had to fit and cement the pieces together as he had to do previously with the Hylaeosaurus and then clear the matrix of bones.

Ascending from the entrance hall to the first story, in the most commanding position of the largest room, were the bones of the Iguanodon, Megalosaurus and Hylaeosaurus. A bust of the Earl of Egremont was prominently displayed along with portraits of his friends, Lyell and Silliman. New expensive cabinets had been introduced. A long case was filled with relics of plants, turtles, fish, birds, shells and Iguanodon tibia from Tilgate forest to illustrate the country of the Iguanodon. Cases along the façade wall displayed fishes and chalk fossils and a mammoth femur. Against the right-hand wall adjoining the fireplace local fossils vied with skeletons of recent acquisitions and donated species from all over the world. Miscellaneous antiquities and diagrams appeared above the case while drawers, not open to the public, held still further specimens for the cognoscenti.

In addition to the numerous academic distinctions conferred upon him, Mantell received from Professor Silliman an honorary LLD (Doctor of Laws) from Yale. Not only had Professor Silliman taken the hint and obtained the LLD diploma for Mantell but declared to Mantell that very few honorary diplomas were granted: 'in the present session – yours and two others (visiting clergymen given the DD (Doctor of Divinity))'. This was very gratifying to Mantell, for although he had practically asked his friend to secure the distinction, and shown his wish for the 'success' of Yale College by a gift[3], yet he could hardly have held great hope of such recognition. Professor Buckland dubbed him the Wizard of the Weald and the Geological Society conferred on him its highest award, the Wollaston Gold Medal. The Maidstone Iguanodon attracted huge crowds and Lyell commented that the collection is a monument of original research and talent, an assemblage of treasures which the mere industry of a collector could never have brought together. It required his zeal, inspired by genius, to bring this to light and call into existence those huge Saurians. For people far and wide the collection was a touchstone of the intellect of the age.

As regards the portrait by a young American artist, which Silliman sent in return at Mantell's request, he replied:

> Thank you very sincerely for your portrait, for although Mrs Mantell has entered her protest against it, and you do not allow it to be a good likeness, and Mr Robert Bakewell Jnr. says the same, it is very interesting to us.

Mantell had seriously miscalculated the essentials of his move to Brighton. From acquiring the lease of No. 20 The Steine in November 1833 to mid-January 1834 everything had been set aside to enable a very

Professor Benjamin Silliman (1779-1864) by George T. Fisher, New York, 1866.

personal and time-consuming involvement in re-establishing the museum, buying expensive new cabinets, rearranging the items and upgrading the catalogue. With the family moving before Christmas the stress must have been considerable with no time given over to his patients. His vanity and pride clearly underlined his mistakes. 'He was extremely agreeable in manner to all who manifested an admiration of his genius'.

Earlier he had considered obtaining an honorary MD so that he could present his medical credentials as Dr Mantell. He had apparently not sought to share a partnership in an existing practice, but assumed that the gentry and even royalty would naturally gravitate to him. He believed that his fame as a scientific discoverer, and a geologist of international repute, would strengthen – not reduce – his professional appeal. He did not endear himself to other medical men in Brighton with a card which at the time was not considered unprofessional:

> Mr Mantell having been engaged in extensive practice in Lewes and its vicinity for upwards of twenty years has been induced at the suggestion of many distinguished families in Brighton who have long honoured him with their professional confidence, to remove to his present residence where he will in future practice the various branches of his profession.

He described himself as Surgeon, Accoucher, Oculist, etc., FRS, and made use of the Honorary LLD to describe himself as doctor. Nor did his literary friend, Horace Smith, help his cause with a poem published in the *Brighton Gazette*, 6 March 1834, of some fifteen stanzas:

> Columbus of the subterranean world,
> Star of Geology, whose rays enlighten,
> What Nature to her darkest depths has hurled
> Mantell! We proudly welcome thee to Brighton

In March Professor Silliman sent a letter introducing Mr Henry Barnard from Hartford, a lawyer, trained at Yale, and now visiting Europe. They met on a day when Mantell was exceedingly unwell but was engrossed to hear about the 'beautiful city' of Yale. That evening they went on a moonlit ramble on the Chain Pier. Mantell described the scene:

> Which in the cool of a summer's evening is a favourite haunt of mine. We sat there till nearly eleven o'clock gossiping about our dear American friends and repeating some of your favourite poets – not forgetting that splendid, and at that hour so peculiarly appropriate piece of Bryant's[4]
> "Spirit that breakest through my lattice – thou
> That cool'st the twilight of the sultry day"

The day after Mr Barnard's departure Mantell was seized with a severe attack of 'internal inflammation', and he records that he was in imminent danger for some hours.

Mantell also miscalculated that his museum would provide a means of support until his medical practice was re-established. However, then as now, professional men were not expected to indulge in business. His professional and scientific friends persuaded him that such obvious proprietorship would damage his reputation and profits as a surgeon; that science should be cultivated for its own sake. By endeavouring to please everyone, he stood to please no one.

Besides establishing his museum with all its attendant publicity and a fresh catalogue, Mantell gave two lectures on geology in February (1834) firstly in the card room of the Old Ship (Brighton's best known hotel) with all the proceeds going to Sussex County Hospital thus emphasising his dual medical and scientific qualifications. Though he read the lecture too fast everyone was impressed with the style of his impassioned and brilliant eloquence. A week later at the National Schools in Church Street he repeated his lecture with less stress to an audience of over 400. Utilising a variety of specimens and diagrams, he estimated the colossal size of the Iguanondon and the florid habitat in which it dwelled. Soon his museum was visited by upwards of 1,000 persons. All the principal people in Brighton had called upon him and inviting Mantell back to their houses. Among the many hundreds of acquaintances, he believed he could rank some real friends.

Following his initial success, the *Brighton Gazette* was keen to promote a course of public geology lectures. Mantell declined alleging professional obligations, realising that he had arrived in Brighton as part of a glut of medical practitioners and in need of establishing his place in the medical hierarchy. This was not to be. He was soon to complain that his practice here was 'very unpromising – hosts of visitors but no patients'.

He absented himself from the British Association meeting in Edinburgh that year, remaining at his post 'as I am resolved to do, that if I do not succeed it may not be my own fault'.

In May he wrote:

> that four months have passed away since I took up my residence here, and yet so eternally have I been engaged with visits and visiting, and journeys to London and Lewes and all the etc., etc.s, of winding up an affair like that in Lewes, that I have scarcely had breathing time and not one moment of quiet that I could devote to the solitude of my own thoughts. Even this day, though but very few professional engagements demand my attentions, I am pressed for time and here not one of what I call my own. My reception in this town has certainly been very flattering so far as visitors and visitings have been concerned but my professional prospects are not encouraging; still I have no right to complain at present, tire and exertion will yet do much for me, and my noble friend, Lord Egremont, whose liberality has pleased me beyond all immediate want of money, having given me a *thousand pounds,* he still countenances me in the most flattering manner.

As ever he committed his thoughts to verse:

> At Lewes I believed the alliance
> Of medicine, surgery and science
> But the kind friends I've found in Brighton
> Have tried my ignorance to enlighten
> And taught me that the kindred three
> Can never have united be
> For the two first have long been made
> Merely a money making trade!
> And of the last that ev'n a grain
> Among them all you'd seek in vain.

The next year he was writing to Silliman, that far from being a fashionable Brighton doctor with clients in the highest circle he now found that the prospect of financial ruin was beginning to form an ugly shape in his mind: 'So here I am, confessedly one of the most successful practitioners in the country with more reputation as a man of science than I deserve and yet without a patient.'

His standing as a surgeon and general practitioner in Lewes strongly supports his claim. He wondered, a view shared with Silliman, that people were suspicious of signing up to a doctor who, it was said, was more committed to geology than to medicine and had no time for his patients. He feared that the rumour was started by rival practitioners. As Professor Silliman replied: 'I am truly grieved that you are so much disappointed professionally in Brighton, but can easily understand how envious rivals may make use of your zeal and of your success too in science to excite prejudice'.

The explanation of his lack of professional success did not require the spreading of rumour, and is better explained by poverty of psychological understanding. The sight of hordes of people of all walks of life flocking to his museum, the scientists and upper classes with their chaises and their horses tethered to the railings of the gardens of Old Steine, would not have encouraged a patient with, for example, a septic finger to venture through the door. To what part of the house would they be expected to go? Most patients desire a degree of privacy when attending the doctor. They are careful who they bump into on the way. As far as the 'nimbies' whose houses abutted Old Steine were concerned, after the initial excitement of the clutter around No. 20 The Steine, they would have felt a reaction similar to that of a modern nimby to the opening an oriental take-away next door.

His own reaction was typical of the man:

> Whatso'er my fate I'll not complain
> If that my humble name enroll'd shall be
> Among the glorious intellectual train
> Whose fame shall live thro' all futurity
> If that the wreath be mine which science twines
> For those who count her smiles and worship at her shrines.

It would not be fantasy to depict the change in his lifestyle with the move from Lewes to Brighton. As a medical practitioner his meal-times may have been irregular but he received gifts of game from the estates of his patients. He would dine as their guest, or with his friends in Lewes. Those who spent the evening with him discussing his collection would be fed; and others would stay overnight. It would be surprising if his housekeeper, supervised by his wife, did not provide good food for the occasion. In Brighton when he was setting up his museum food would have to be brought up several flights of stairs. It would tend to be cold and less fresh than in Lewes. Working, and probably living, in the basement the servants would be more tetchy. They would have to carry buckets of water up several stories for ablutions and to clean his 'stones' i.e. fossils. They would grumble. The coachman would probably live a little way away, near the horse and chaise. If he doubled up as footman as in Lewes this would be more difficult.

In Lewes his wife was prepared to show some visitors around the collection; here she would fight for space in the upper rooms which would also serve as his study and an area were he worked on his fossils before exhibiting them. His patients would have to ascend the stairway above the museum to be examined; whereas other social doctors would bring their patients into a desirable ground floor waiting room next to an equally opulent consulting room.

With irritation all around, with poorer food, the absence of his family, the dissatisfaction of his medical practice and the large number of visitors to his museum, his physical as well as his mental health would have been that much the poorer. He was running out of money and the future was uncertain. At this stage he wrote to his friends wondering whether he should take the opportunity to go on a voyage of discovery as surgeon and naturalist or become a lecturer in Physiology and Geology, either in London or Philadelphia. He rejected these ideas because of the uncertain health of his younger daughter, Hannah, and because his elder son, Walter, was likely to enter the medical profession. He also contemplated writing a book on the nervous system.

The season ended in Brighton with the court's departure in February leaving the town comparatively empty. None of his social or scientific triumphs brought him any money and his expected practice failed to materialise.

Later that year he delivered the first geological outdoor lecture in Sussex. The excursion was along the shore to Rottingdean and from thence to Castle Hill, Newhaven, where there was a marquee and a buffet. The day was cloudless and charming, Mrs Wagner, Mr James Anderson, Mr Horace Smith, Ricardo, Masquerier, in all forty persons including nine or ten ladies, were in the party. He had made himself a most attractive lecturer, filling the listening ears of his audience with seductive imagery and leaving them in amazement of his exhaustive catalogue of wonders. His style, both in speaking and writing, was fluent and singularly free from technicalities – qualities which went far to render his works popular, and make him a favourite among the various institutions for diffusing knowledge among the people.

Afterwards he stated that he dare not organise any further excursion for 'would you believe it certain detractors have attempted to ruin my professional prospects by asserting that I devote all my attention to geology'.

On Wednesday 29 October 1834 Agassiz (the Swiss Geologist), Buckland, Lyell, Moses Ricardo, Robert Bakewell and Michael Faraday all visited the museum. Over the next few days he dined with Buckland, Agassiz, Lyell and Bakeweil; but on the 31st his wife became very ill, a fact which he resented as it kept him from enjoying his distinguished guests. He complained in his journal: 'Mrs Mantell was very ill which sadly interrupted my intercourse with my friends and dampened the pleasure I should have had in their society'.

Louis Agassiz (1807-73) was a Swiss naturalist who when young took up the study of living fish. Moving to Paris at the age of twenty-four, he met Cuvier and took over and extended his collection of fossil fish, returning to Switzerland as professor at Neuchatel. He settled in America in the late 1840s, building the original wing of Harvard's Museum of Comparative Zoology. Mantell was justifiably proud of his fossil fishes from the Chalk and Wealden and utilised the *Brighton Gazette* to highlight

> The skill and care with which this collection had been made, and the admirable manner in which Dr Mantell has dissected the specimens, removing the surrounding rock from the preserved parts of the animal ... From the result of M. Agassiz' examination of these remains it appears that the chalk of the SE of England contains 23 species of fishes, of these 18 belong to extinct genera; 14 of these fishes have not been found elsewhere.

One remarkably large fish with spinous scales was named by M. Agassiz as Macropoma mantelli. The fourth livraison of Agassiz' *Recherches sur les Poisons Fossils* contained extensive tabulation of Mantell's chalk fishes and Agassiz returned to Brighton in October to complete arrangement with his artist Joseph Dinkel whose drawings of Mantell's chalk fishes appeared prominently in atlas segments four and five of the book. Dinkel befriended Mantell and his daughter Ellen, giving her drawing lessons that would determine the course of her life.

In April 1835, his demeanour seems to have improved:

> I have had neither the inclination nor the leisure to open my notebook. Since my residence in Brighton for eighteen months the following pleasing events have occurred:
>
> 1. A piece of plate presented to me by some of my Lewes patients,
> 2. Two lectures, received with great approbation,
> 3. An outdoor lecture from Brighton to Newhaven,
> 4. The Earl of Egremont presented me with £1,000
> 5. A fine specimen of Iguanondon presented to me,
> 6. LLD,
> 7. A work by Dr Morton of Philadelphia on the Organic Remains of the secondary formations of the United States dedicated to me by the author,
> 8. The Wollaston award with 10 guineas (a gold medal) and £22 in money from the Council of the Geological Society.
> 9. Mr Lyell at the Anniversary of the Geological Society in a very eloquent speech gave an account of my researches,
> 10. Received a specimen of icththyosaurus tenuirostris from Lord Cole and a fossil tree from the Isle of Portland.

Any euphoria did not persist. Mrs Manrtell remained in London from April returning on 2 June. On 25 April he complained of being still very unwell. 'I believe I shall never regain my strength. My practice is very disappointing.' Three aspects of his life left him floating on a sea of circumstances: financial difficulties, marital and domestic affairs and his state of health. Throughout much of 1835 he remained sick, listless and despondent, unable to find the path of peace. He became provocatively contentious having minor tiffs with his scientific colleagues and even his Brighton friends. As stated in July the day after sitting late into the night at the Chain Pier with a friend he was seized with an attack of 'internal inflammation'. Spokes assumed that he had a chill and wonders whether he was being hypochondriacal. But he had further trouble in August with cold rigors lasting several hours succeeded by fever and delirium which left him in reduced state when compelled to visit several patients. He remained in ill-health for the rest of the year developing a swollen face in September. He attended a pantomime with his children on 31 December, a day when several gentlemen arrived at his house to draw up formal proposals to turn his home into an institution.

His ambition and his domineering personality played havoc with his family life. Initially Mary had been anxious to help him with his geological adventures, visiting the quarries with him and drawing the fossils for his publications. He did not neglect his family but as the workload increased, both medical and geological, his interests caused him to overlook his wife's health; he no longer empathised with her needs, failed to understand her emotions and lost interest in her psychological problems.

She went away for weeks to recuperate from illness. In Lewes she often showed visitors around but found the house increasingly cramped for space. The situation was even worse in Brighton when he was also frustrated by lack of progress medically. Even when the children were away at school the domestic situation was increasingly cramped. Cadbury highlights Christmas 1835 when events reached a vortex failing to speak to each other, despite the fact that she was often seen with one of the children attending his lectures.

His professional prospects continued to be unfavourable although he had carefully abstained from all public exertions in the cause of science. This had the effect of inducing some of his friends to attempt the formation of a Scientific Institute based on the public exhibition of his collection from the loan of which they were to allocate to him £200 per annum. Meanwhile Mantell had thoughts of placing his family in a little cottage near London and offering his services as surgeon or naturalist to any expedition going on a voyage of discovery.

On 26 December 1835 he lectured on 'Corals and the Animals which form them'. Presumably this was based on what he had learnt from Charles Lyell but it is also the first lecture he gave based upon microscopic observations. He marvelled at the organisation of beings so miniscule as to be invisible to the naked eye and their mode of life and action on the important physical changes effected by such apparently inadequate agents.

On 1 January 1836 he met with the subcommittee formed to inaugurate the Sussex Scientific Institution and Mantellian Museum. Their Majesties had stated that they could not allow their names to be associated with the project. 'So much,' Mantell remarked, 'for the advancement of science and the encouragements it gives to its cultivators.'

On 4 January Sir Richard Hunter, Moses Ricardo and Horace Smith visited Lord Egremont at Petworth to request his name as Patron. Lord Egremont with his accustomed liberality gave £1,000 for the project. Davies Gilbert was appointed President and George Fleming Richardson the Curator.

In many ways the appointment of G. F. Richardson as curator of the museum was an excellent choice. He was the brilliant son of a silk merchant but was unwilling to follow his father's trade and instead devoted his time to the study of classical and modern languages, Greek, Latin, Hebrew, Italian, French and German. He had a prodigious memory and incredible aptitude for acquiring knowledge. Thus by the time the Scientific Institution opened he had familiarised himself with geology. Soon he was able to lecture not only on the language and literature of Germany; and on the ruins of Lewes Priory, but on geology along with Mantell. He translated Hermann von Meyeer's *Palaeologica* from the German in which the Megalosaurus and Iguanodon were paired as Saurians with limbs similar to those of heavy land mammalian. Edward Charlesworth published the English version with a forword by Mantell who appended a description of Hylaeosaurus. In *The Wonders of Geology* Mantell added that it was based on a series of lectures which he had given from notes taken by G. F. Richardson. Richardson was able to help in other ways adding a poem to Mantell's *A Day's Ramble in and about the Ancient Town of Lewes* and sending a poem of condolence after the death of Mantell's younger daughter.

These events were a turning point in his relationship with his wife. Letting the house, at a financial loss, meant that only a parlour and a bedroom was left for himself from which he expected to carry out his professional work. For her, the plight they were in was '*all his fault*'. It was his reckless pride and his selfish pursuit of his own interests. He saw nothing but ruin before him. It was a sad trial for him and the more so as the '*sacrifice*' was not appreciated. In April his family moved to a cottage in Southover Street, Lewes, near to Reginald's school, at the foot of the hill below Castle Place which he had failed to sell. Spokes comments: 'that one cannot help feeling that the position of his family on their return to Lewes was not such as to be exhilarating, or to make their sojourn there enjoyable.'

The following autumn he gave a further series of geological lectures. In March 1836 he lectured at the Old Ship 'On the Extraordinary Fossil Remains of a Crocodile (Goniopholis) recently discovered at Swanage' that Robert Trotter, the generous 'poacher' of Borde Hill had purchased from quarrymen in Dorset and presented to the speaker. In the same month he visited his fourteen-year-old daughter Hannah Matilda at her school in Dulwich to find that she was severely ill: 'she was attacked

Davies Gilbert (1767-1839) a friend, neighbour of Mantell, and later President of the Royal Society.

with disease of the hip joint while at school and unfortunately the earliest symptoms of that infection were mistaken for a mere common rheumatism'. Furthermore his brother Joshua suffered an accident with permanent brain damage, apparently the result of a fall from his horse. The family returned to Brighton that autumn and occupied a small cottage in the western part of Brighton (possibly Hove) but he retained his rooms at the Steine for his practice and the arrangement could lessen the wreck of his little fortune. He resolved to stay through the next season and then try a practice in London.

In the spring of 1836 in a public lecture, his son Walter spoke on geology while he discussed local antiquities such as he had studied with the Rev. James Douglas. In the autumn he gave three public lectures on physiology at Brighton Town Hall which were reported with acclaim in the medical journal, *The Lancet*. In May he went to Lewes with Mrs Mantell and a few days later met Mr Davies Gilbert at Mrs Newton's in Southover and many members of the institution and their friends. Then, at two o'clock in the afternoon, they assembled at Southover church, in all about sixty persons in twenty carriages, one-third ladies. Mantell described the Priory and gave a brief sketch of the history including Mount Calvary and Gundrads's tomb. They walked over the hill to the cliffs, where they took to their carriages and drove to the Castle banks later visiting St John's church to inspect the monument of Magnus. The party then adjourned to the Castle where a cold collation was prepared in a fine marquee; about sixty at table. Mr Davies Gilbert took the chair and everything passed off very agreeably, notwithstanding the cold and damp weather.

In August he went to London to see Lady Mantell and accompanied her to the Surrey Zoological Gardens to visit the giraffes, lately brought over – 'very fine creatures' – one of them fifteen feet in height. On the 16th of that month Lady Byron (widow of the noble Poet) accompanied by a lady, came to the museum, and spent nearly two hours in examining the chalk fishes, etc. He presented Lady Byron with an elegantly bound copy of the *Geology of SE of England*, describing her 'as a woman of delicate form, rather pallid, mild and very lady-like in her manners and conversation – she appeared deeply interested in the subjects of our conversation which was confined to Geology and Antiquary'.

The winter of 1836-7 was bitterly harsh. A snowstorm caused drifts which made the roads impassable. An avalanche at Lewes killed thirteen or fourteen people, many of whom had been his patients. Even in April the snow was still on the grass and the cattle suffered accordingly. In May 1837 the weather had eased and he lectured at the Royal Institution in London on 'The Iguanodon and other Fossil Remains discovered in the Strata of the Tilgate Forest'. A total of 700 people were present and afterwards he drank tea with Lord Northampton. He gave a further geological lecture in Brighton in September when only five people were present but was much cheered by a visit in July to the Devil's Dyke on the crest of the South Downs. He arrived there at 7.00 a.m.

> The most glorious sight imaginable – the sun breaking through a mass of clouds poured streams of living light on the landscape – the distant downs, by Steyning and to the far west were crested with mist and the reflections of the sun's rays, gave them the magical effect which is seen on the snow-clad Alpes. This gorgeous scene continued about ten minutes and then all was wrapt in gloom. Broke the spring of the carriage and obliged to walk home.

His younger son, Reginald Neville Mantell, was a schoolboy aged ten who spent his spare time examining the sand on the seashore and his father decided to dedicate to him a little gilt-edged book of forty pages, *Thoughts on a Pebble*, intended to be the first lesson in geology. It ran through many editions, the seventh in 1846. As he was later to write: 'My Pebble is doing much good, for it winds its way where scientific works would not be permitted to enter many a young mind'.

From the latter period of his time in Lewes, Mantell's concern with those less well off in society gave way to his self-interest in maintaining his practice and establishing his name with his museum, lectures and membership of various societies. Others recognised the change. Bakewell wrote:

> My good friend how comes it to pass that you say not a word about reform. Have Sir John and Lady Shelley transported you into a Tory. Let me recommend you to get the jaw of a Tory into your museum, for they will soon be extinct animals.

Certainly there is little mention in his diary of the Reform Act, Charterism[5], the Poor Law and Anti-Corn Law Acts; instead he objects strongly to the imposition of the new income tax. Income tax was originally introduced in 1798 to finance the war against the French and was duly abolished in 1816 when that emergency was over. Peel reintroduced Income Tax in 1842 at the rate of 7*d* in the pound on incomes over £159. In theory it always remained a temporary measure.

Both Mantell and the museum's curator, George Richardson, felt strongly that the fate of their new Scientific Institute in Brighton depended on the success or failure of their attempts to cultivate a taste for scientific knowledge in the town. Mantell's talks became increasingly popular with up to 800 attending along with distinguished visitors such as Michael Faraday, Buckland, Murchison, and Lyell. By now it was the Sussex Royal Institution – the king and queen having affixed their names as patrons, though neither the queen nor the younger members of her family had been to see it. By July 1836 some form of family reconciliation had occurred and he wrote optimistically: 'Happiness may yet be my lot of those I love. My sweet Hannah Matilda decidedly better!'

Three weeks after a highly successful lecture in October 1837, Lord Egremont died; the planned anniversary dinner gave way to a hastily arranged emergency meeting which spelt the death knell for the museum. It was offered to Brighton Council for £3,000 but they turned down the offer. The only other possibility was sale of the many thousand specimens to the British Museum. In January 1838 he wrote to Charles Konig:
'My dear Sir, I am desirous of entering into a treaty with the Trustees of the British Museum.' And in his application of 17 February he stated:

> The Collection consists of many thousand specimens but the grand features and on which I rest the claims to your attention are the remains of the Iguanodon, Hylaeosaurus and other colossal reptiles and fossils peculiar to the Weald of the South East of England on the Chalk fossils of the South Downs, of the latter I will beg to observe that the last livraison of M. Agassiz contains 20 folio plates devoted solely to the fishes of my collection. The sum for which I offer my Museum to the National Collection is £5,000 which is much less than I have expended in the formation and their marketable value according to the opinion of some eminent savants and competent judges. I am anxious that this collection which so fully illustrates the physical structure and ancient condition of one of he most interesting and peculiar districts of Great Britain should not be sent out of the United Kingdom that I am willing to enter into any arrangement and negotiation which you, my Lords and Gentlemen, may purpose should you be pleased to entertain a favourable opinion of my application.

The government declined to purchase his collection on account of the defalcation – misuse, misappropriation or inadequate collection – of the revenue which had pitched the country's finances into turmoil, but some friends led by Lord Northampton remonstrated with the Chancellor. By the end of July the Trustees opted to give £4,000 and undertake the packing and removal at their expense[6]. In effect the total cost was nearly equivalent to the sum he had originally intended. However, 'John Bull being too poor to pay me this year', payment was delayed and he had to wait a year for the money.

He was greatly concerned about the packing of the specimens and suggested that proper cases should be prepared for the Maidstone Iguanodon, the Hylaeosaurus and the Swansea crocodile. His preferred plan was the removal by sea from the chain pier with a stately entrance to London by the Thames. This was not to be. The removal was capably orchestrated by Konig and Richardson. He awoke on 10 December to the sound of horses. In the mayhem below there were ninety horse-drawn vans ordered to take the entire collection.

Horace Smith noted the chaos of the removal of Dr Mantell's Collection from Brighton to the British Museum in London:

> Doctor Mantell's Museum was all disarranged
> The huge fossil bones on the carpet were laid
> To be packed for conveyance to London
> The curator was gone – there was no other guard
> All the drawers were unlocked, all the doors were unbarred
> All the cases were opened and undone.
>
> In the midst of these relics, o'ertaken by sleep
> A dream whisked me suddenly back at a leap.
>
> Amazement soon yielded to terror – for lo!
> The bones of the grand 'gan to rush to and fro,
> And to form into skeletons antic,
> Each quickly its private carcass indeed
> Till the monsters all starting to life, were renewed
> In their former dimensions gigantic.
>
> As they rose to their height, they uplifted the roof
> While the beams of the floor, 'neath each ponderous hoof
> Like so many lathes split asunder
>
> And forth from their prison – as fierce as when first
> They ravaged the forests and waters – they burst
> With a rush and the roaring of thunder.

The loss of his collection was a sad moment, as his diary recalls:

> What a lesson of humility; what a proof of the vanity of human expectation; but so it is, and so it ever will be, yet I do not regret that I have spent my days and nights and toiled in vain!

SCIENCE

SCIENCE, PROGRESS AND ENLIGHTENMENT

> *To insist on the high importance of Scientific Studies, whether as enlarging the sphere of our intellectual enjoyment, or as contributing to our rank and power as a nation, would in the present age be altogether superfluous.*
>
> (*Quarterly Review*, 1830.)

Gideon Mantell's life was fired by a restless energy with ambition to match, determined to achieve immortality in the realms of science. Thus, in many ways he personified the age in which he lived. Science and innovation led to the Industrial Revolution which started in Britain. An American historian, Robert C. Allen, (*The British Industrial Revolution in Global Perspective*, 2009) argues that the Industrial Revolution took place in Britain ahead of the rest of Europe because, in the early 1700s, Britain was a country of conspicuously high wages (dating from the aftermath of the Black Death), cheap sources of energy, and flourishing overseas trade. Its output in the mid-nineteenth century outstripped Europe and America combined. As farming became more productive, people were able to move into the cities helped by the provision of cheap fuel. Britain had a thriving capital market and this added an edge to its industrialising.

In an examination of some of the early men of science, J. Uglow (*The Lunar Men*, 2002) believes of Dissenters that only a few of them had a university education and most were free thinkers 'united by a common love of science which was thought sufficient to bring together persons of all distinction, Christians, Jews, Mohametans, and Heathens, Monarchists and Republicans' and provide the impetus for progress. After the 1660s when the punitive laws of the Clarendon Code banned them from worshipping in the chartered towns and the Test Acts of 1675 excluded Dissenters from public office, teaching and the universities, they made use of their talents, unhampered by old traditions of deference and stuffy institutions, to experiment and establish a social conscience. Ironically this exclusion from the old universities, where the Classics-based courses had hardly altered since Tudor times, was a great spur to British culture. After studying elsewhere they set up Dissenting Academies positively welcoming the new: teaching of modern languages, modern history, mathematics, politics and natural philosophy. This approach stemming from the Age of Reason and Enlightenment not only accords with Mantell's behaviour, but became widespread throughout Britain. People experimented, collected and exchanged ideas.

The growth of the scientific community led to a more questioning and critical appraisal of the established order: John Playfair and Charles Lyell had criticised the universities for failing to recognise and reward the cultivators of English science and Charles Babbage in his *Reflections* of 1820 attacked the complacency of the Royal Society. Sir Joseph Banks (1743-1829) a botanist who had sailed with Captain Cook in the *Endeavour* and visited Newfoundland, Labrador, Iceland and the Scottish isles before advising George III in 1771 to set up Kew Gardens, presided over a long, tyrannical reign of forty-one years which ended with his resignation a month before his death in 1820. More kindly described as a courtly regime, the Royal Society was assailed from several directions, uneasily sharing premises at Somerset House with the Astrological and Geological Societies; nonetheless it came to be 'the most successful networking society' extending abroad and influencing the machinery of government. Sir Humphry Davy was elected President in 1820 and though he mended some bridges was largely ineffectual, resisting the forces of change, criticised for his choice of authors for the Bridgewater Treaties on Natural Theology, and accused by Charles Babbage and others of 'borough mongering'.

Those ranking among the ruling classes believed they had a constitutional and hereditary right to preside over the burgeoning societies. Societies also attracted those with leisure and interests varying from the superficial to the profound in science, literature and the arts. But most importantly there was an upward thrust of self-advancement among those, as Thackray declared:

> Disdaining class barriers and espousing progressivist values of Mankind and a progressivist interpretation of science which can be seen as deriving from their need to justify themselves to do so in terms of belief systems that simultaneously affirmed their commitments to high culture, announced their distance from the traditional value systems of English

> Society, and offered a coherent explanatory system for the unprecedented and change oriented society to which they found themselves unavoidably if willingly cast in a leading role.

The Geological Society was formed in 1807, initially as a dining club but rivalry between the Royal Society and the Geological Society erupted in 1808 when Sir Joseph Banks and Humphry Davy, with others, tried unsuccessfully to control the functions of the Geological Society. Having failed, they resigned from the Geological Society in 1809 with many claiming that the pursuit of science does not constitute a distinct profession. Papers such as that by Mantell on the Iguanodon continued to be read and published by the Royal Society but there was a risk that interest in science might lessen. Furthermore, minor disputes may have involved differences in personalities: Anglican versus Dissident; evolutionist versus anti-evolutionist. For these reasons the British Association for the Advancement of Science (BAAS) was founded in 1831 as another rival to the Royal Society, giving a boost to science in the form of an itinerant association, meeting annually over several days at provincial centres. At the Plymouth meeting in 1841, Robert Owen gave his controversial paper on Dinosaurs. His paper and the subsequent report led Mantell to reply as he had previously done via the Royal Society, provoking Sir Robert Murchison, the President of the Geological Society to take umbrage, commenting:

> Whilst I understand the motive which led Mantell to communicate his last memoir on the Iguanodon to the same Society that he had addressed his first account of the saurian, I regret that he should not have communicated to ourselves other palaeontological remains.

It was probably for this reason that he did not visit Mantell when Mantell moved to central London.

It is not possible to assess precisely to what extent Mantell was a spectator, affected or embroiled in the disputes and manoeuvrings of the Royal Society; but he was intimately associated with the principal characters of the drama. The major player in manipulating the policies of the Royal Society was Davies Giddy, a mineralogist, formerly MP for Bodmin and Helston, and at one time High Sheriff of Cornwall. On his marriage he inherited an estate in Sussex and changed his name to Gilbert. Mantell's *Illustrations of the Geology of Sussex* was dedicated to Davies Gilbert, who later became President of the Mantellian Museum in Brighton, and on 10 February 1825, read a letter from Mantell before the Royal Society entitled 'Notes on the Iguanodon, a newly discovered fossil reptile from the sandstone of Tilgate forest in Sussex.'

Davies Gilbert had befriended Thomas Paine in London and was elected to the Royal Society in 1791 having been sponsored by his geological neighbour and fellow radical, John Hawkins, later of Bignor Park, Sussex. From 1806 he cast off his radicalism and became an admirer of Fox and Burke. He warned parliament that the tumultuous storm of democracy would lead to a gulf of despotism, spoke for public order in the face of the Peterloo massacre and was against Catholic Emancipation. He became a prominent member of the Linnaean and Geological societies. When Sir Humphry Davy retired in 1827 Gilbert was invited to become President but instead asked Sir Robert Peel to stand. When Peel declined the invitation, Gilbert was elected at the age of sixty. Within a month he squashed the report of the Fellowship Committee on the Society's constitution and Herschel, then aged thirty-five, resigned as secretary. After eighteen months Gilbert stood down in favour of his nominee, the Royal Duke of Sussex. Herschel stood in opposition to the establishment candidate, supported by Babbage, Peacock and Whewell, losing on St Andrew's Day 1830 by 119 votes to 111 in favour of Sussex. Mantell in his journal expresses surprise that the Duke should have accepted the position in view of his narrow majority but gives no indication where his sympathies lay or even whether he personally voted.

For the next eight years, Sussex, aided by Gilbert, tried to create a rapprochement between the Royal Society and the political life of the country. With his combination of political wisdom and intellectual taste he began to cultivate the Fellowship with the minimum of change. Herschel left the country to begin his observations at the Cape of Good Hope. It was ostensibly at Sussex's request that Herschel was given a baronetcy in 1838. Babbage, Murchison and Sedgwick turned to the new British Association for the Advancement of Science leading to an expansion of scientific activity, characterised by large scale provincial meetings with days of receptions, scientific discussion and explorations. In a change of direction the Royal Society worked through the BAAS for government support for tidal observation, Antarctic expeditions and magnetic observation. However, in other respects the Royal Society in 1840 was neither completely dignified nor efficient. It was said that the focus was on aristocratic science and scientific aristocracy.

In 1838 Sussex welcomed the Council's nomination of the Marquess of Northampton as his successor; a man, according to Lytton, without more pretensions to science or philosophy other than a good education and a taste for such studies would justify. At forty-eight he had been a Whig MP (1812-20), an FRS and a mineralogist of note. Under his direction the Royal Society completed the third and final instalment of its internal structure. In 1848, under the Earl Rosse, an element of Tory paternalism prevailed in the Society's government.

Both in London and in the provinces societies burgeoned forth: the Royal Society (received its charter dating from 1662), the Linnaean Society (1788), the Geological Society (1807), the Astronomical Society (1830), the Zoological Society (1826), and the Chemical Society (1841). The Royal Institution (founded in 1799), could hold an audience of 800, hear the music of Handel and Haydn and the lectures of Davy and Faraday. Everywhere people sought knowledge tinged with excitement. People read. Books became more readily obtainable, though often printed with difficulty and dependent on subscriptions before production. Books were available from libraries or if bought were passed from hand to hand. News-sheets, gazettes and other local papers were printed and distributed. To be seen to be literate and to possess knowledge gained importance. More and more people became aware of the world around them: of foreign wars, the behaviour of the royal family, the decisions of the law courts, markets and what was locally available. These matters were discussed. It was a continuation of a surge of intellectual activity which began in the last decades of the eighteenth century. A great deal of the momentum stemmed from moral reforms and philanthropic societies often linked to various churches and denominations. Clubs and societies were among the most numerous, diverse and dynamic organisations of Georgian society challenging the drinking houses in their level of support. The field of ideas emanating from the revolutions of that time were paltry in comparison with the scientific discoveries and technical inventions which swept through the land.

There was debate. The same was also true concerning books on the natural world. The layman could become interested and involved, but science became embroiled in religious controversy. To quote Charles Lyell, 'England was more parson ridden than any country in Europe except Spain and this was detrimental to scientific thinking'. How literally could the Book of Genesis be interpreted? What were fossils? Had there been more than one deluge? Was there an age of reptiles before man and mammals? Did they live on land or in water? Scientists debated among themselves, aware of public interest. Geology captured the public imagination. Novels such as Jules Verne's *Journey to the Centre of the Earth* – a geological epic based on what was known of the earth's interior – abounded. Even the astronomer, Sir John Herschel, enthused that geology in the magnitude and sublimity of the objects of which it treats, ranks next to astronomy in the scale of the sciences. When the British Association for the Advancement of Science met in 1845 geology was allocated the largest and best room of any of the sections, providing a social pageant. It was the most numerously attended and the only section enjoying feminine patronage mingling more pleasantries with science than any of the rest.

In 1843 Mantell, recovering from his paralysis, could no longer rely on his medical practice or fulfil his promise to Lyell to concentrate on discovery of new fossil species. He applied himself to developing popular works on geology to satisfy his pecuniary requirements. To do this he employed a narrative style combined with an accurate adherence to scientific facts. He spent much of the year preparing his *Medals of Creation or First Lessons in Geology and in the Study of Organic Remains*, writhing with severe pain at the task. There were 1,000 pages of text and 100 illustrations: some produced by Joseph Dinkel and others by Ellen Maria Mantell. It was technical, voluminous and exacting, and though designated in part for a popular audience, *Medals* was arguably the first modern synthesis of palaeontological knowledge in English. Henry Bohn, the publisher planned to produce it in two volumes at a cost of one guinea. In the first instance he would publish 2,000 copies from which Mantell would not recover the costs save reimbursement for the illustrations. The project developed slowly and by July 1844 only 500 copies were printed. It featured a series of geological excursions as for example around Matlock in the Peak District. Travellers were advised to bring a hammer, leather or camlet bag, a paper for wrapping specimens, boxes, wadding, string, labels, gloves, eye preservers, tape, compass, map, knives, chisels and a set of lenses; and were taught how to collect and how to sketch. After a moralistic introduction as with some of his earlier works, he made a chronological survey of British strata and their fossil contents. He described how by slow and almost insensible gradations the present state of life had come about with an age of mammals, including monkeys, following an age of reptilia. *Medals* would have been disseminated widely: a few wealthy buyers would have a copy for their libraries and where they could afford, lending libraries such as at Mechanics Institutes would have been prepared to buy. Most books invariably passed from hand to hand with a far larger readership than the sales might suggest.

However, despite its success, its publication in 1844 clashed with other popular and lighter publications, including one from his assistant, George F. Richardson, and the anonymously produced *Vestiges of Creation*.

Mantell's early works on the *Fossils of the South Downs* (1822) and *Illustrations of the Geology of Sussex* (1827) drew controversy in so far as he had yet to establish that the Iguanodon was a herbivore and had to dispute whether the Saurians were marine or land reptiles. The first serious antagonism came in 1835 when Sir Richard Owen (1804-92) tried to establish his credentials. Owen grew up in Lancaster. His father was a merchant trading in the West Indies and his French mother, Catherine, founded a girl's boarding school. At sixteen he served an apprenticeship to a surgeon apothecary, like Mantell developing a skill in dissection and knowledge of human anatomy, and discovering his vocation. In 1824 he went to the University of Edinburgh to complete his qualifications but remained only two terms. He avoided the lectures of Alexander Monro, the third generation of professors of anatomy, who was responsible for Charles Darwin's aversion to human anatomy and medicine in 1825. However, he did attend John Barclay's course in anatomy which were generally recognised as far superior to that of the third Monro. From Barclay's earnest teaching he gained a strong predilection for 'Zootomical pursuits' and, on his recommendation, came to London to complete the requirements for MRCS (Master of the Royal College of Surgeons) thus enabling him to pursue private practice. Barclay strongly recommended him to Abernethy. He was still too young to qualify for MRCS but Barclay's reference impressed Abernethy who took him on as prosector for his surgical lectures. The appointment, though unpaid, was fortunate in that it relieved Owen of the financial burden entailed by the need to complete the course requirements and at the same time provided him with valuable experience as a practising anatomist. And he was able to attend J. H. Green's lectures on the comparative anatomy of the whole animal kingdom.

At the age of twenty-two, Owen passed his MRCS and started to practice. Meanwhile the Royal College of Surgeons was charged by Thomas Wakley, in *The Lancet,* with failing to produce a useful catalogue of John Hunter's collection purchased by the government in 1800 and entrusted to the college. William Clift, (1775-1849), Hunter's last student assistant, had come to London from Devon with Walter Raleigh Gilbert, Gentleman to the Bedchamber of George III, and was apprenticed to the Society of Surgeons to provide Hunter with anatomical drawings and secretarial aid. When Hunter died his estate, including his collection, was initially in the hands of Sir Everard Home and Matthew Baillie but Clift copied out half of Hunter's manuscripts and at parliament's behest was employed as conservator of the collection which he had to move on two occasions from Castle Street to Lincoln's Inn Fields and thence into the new college building.

Abernethy engineered Owen's appointment as his assistant. Owen entered the Hunterian Museum of the College on 7 March 1827 at a salary of £30 per quarter. His salary was soon increased to £150 per annum. He maintained a middling medical practice and published significant papers on comparative anatomy, becoming acquainted with the medical establishment and establishing a good rapport with Clift. His fluency in French, taught by his mother, enabled him to escort Cuvier when he visited the Hunterian Museum and later he was invited to Paris by Cuvier; where he worked under supervision as an anatomist for a month. He also became active as the youngest member of a group who found the Zoological Society of London. When Clift's son died in a street accident he became co-conservator with Clift, completing the cataloguing of the Hunterian collection.

In appearance he was thin, becoming more gaunt as he aged. He seemed even taller than his six foot height, with a massive head, lofty forehead, high cheek bones and curiously rounded and prominent eyes. He had a large mouth, projecting chin and long, lank, dark hair. He shared with Clift a taste of the theatre and music, eventually marrying Clift's daughter. Clift had groomed his son to succeed him, but when the son died Owen naturally stepped into his place. In 1836 he was appointed Hunterian professor of comparative anatomy and physiology at the Royal College of Surgeons. As an anatomist he assumed the mantle of Cuvier with the prestige of the Royal College and the British Museum to support his pre-eminence in the scientific societies of the capital. In the field of palaeontology he was considered a disciple of Buckland.

Owen had assiduously followed the advice of his mother:

> One thing that can never be too strongly recommended to a young man aspiring to rise in his profession is to become pupil of some person already eminent and in high repute; by such a course they obtain two great objects: a well grounded professional knowledge and the opportunity of becoming known to all the friends and connections of their instructor.

He was regarded as a brilliant but unscrupulous comparative anatomist, working from his laboratory in the museum of the College of Surgeons; and from 1828 giving lectures in comparative anatomy at St Bartholomew's. His envy of Mantell's original discoveries 'from the strata, the quarries, chiselled, washed and rearranged' (from in situ and in vivo to in vitro) spoilt what should have been a congenial relationship between them. Huxley commented that Richard Owen had many excellent and great qualities and one fatal defect – utter untrustworthiness. 'He is both feared and hated'. He later wrote that it is astonishing with what an intense feeling of hatred Owen is regarded by the majority of his contemporaries – with Mantell as the arch-hater.

> The truth is that he is superior to most and does not conceal that he knows it and it must be confessed that he does some very ill-natured tricks now and then. Owen is an able man but to my mind not so great as he thinks himself.

Huxley was to cross swords with Owen when Owen briefed Bishop Wilberforce for the great debate on evolution. After Owen died Huxley was tasked with summarising Owen's achievements and in a letter to Joseph Hooker commented that 'I am toiling over my chapter about Owen, and I believe his ghost in Hades is grinning over my difficulties.'

Owen was accused of thriving on feuds and antagonisms, and of unworthy piracy and ingratitude. He ignored the work done by Chaning Pearce on Belemnoteuthis. He opposed Mantell's gold medal and attacked him in his obituary. He ridiculed the name Iguanodon. He wrangled with him over an Iguanodon jaw, which probably belonged to another type of saurian. He disputed whether bones found by Mantell in Tilgate forest were those of birds or of a pterodactyl; and appropriated various lithographs which Mantell had collected. Lyell tried to remain outside the controversy until Owen attacked him regarding the extent to which the Book of Genesis should be regarded as historically valid. When Captain Brickenden found the Telerpeton near Elgin and gave it to Lyell who asked Mantell's opinion on its anatomy, Owen, using a dastardly ruse, renamed it Leptopleuron lacertinum.

There were to be other disputes. Though Mantell attended many of Owen's lectures and placed himself as the innocent victim of Owen's greedy ambition, he himself was nonetheless a prickly character when faced with even mild criticism. He was angry with 'poachers' such as George Bax Holmes (the sly Quaker), and Robert Trotter of Borde Hill near Cuckfield who were excited by Mantell's discoveries and were anxious to acquire fossils for themselves from the quarrymen who had supplied Mantell from Tilgate forest, 'poaching' what should have come to Mantell directly. The Reverend Henry Hoper 'poached' from other quarries at Castle Hill and elsewhere. Holmes had sent specimens to Owen but became aggrieved when Owen did not return them but claimed them as his own. Mantell refused to speak with the 'poachers' but all three were later to offer fossils as gifts to Mantell. William Harding Bensted who owned the quarry near Maidstone was not prepared to sell his Iguanodon at the price Mantel proposed and friends clubbed together to buy it for him. Despite this he had reason later to be grateful to Bensted from whom he later purchased the skull, ribs and vertebrae of an ancient turtle.

The commercial success of the anonymously published *Vestiges of the Natural History of Creation* was both a symptom of how fascinating these matters had become to the general public and a cause of a growing obsession with fossils in general. The author remained anonymous to the public at large during his life time. Speculation as to who was the great Unknown included Thackeray, Harriet Martineau, Lyell and even Prince Albert, until in 1884 a posthumous acknowledgement of authorship was published. It is possible that Lyell knew fairly early on and others in the scientific community slowly learnt its origin. In 1852 at a meeting of the Geographical Society Mantell chatted with the author. He is described in the *Dictionary of National Biography* as having six digits on each limb – a hexadactyl; the additional toes on his feet had to be amputated to enable him to wear shoes and take part in sport With just 390 pages the ably named *Vestiges* went through twelve editions and sold 40,000 copies (as well as a very popular pirated edition in America). The author was one Robert Chambers, a Caledonian journalist and clearly not a scientist – a fact which aroused the scorn of the scientific establishment. Among several mistakes, Chambers believed that birds were the ancestors of the duckbill platypus and later mammals; and oats could be converted into rye. He added a series of wholly imaginative genealogies in which, for example, pachyderms were descended from manatees, dogs from seals, otters and polecats; but most of his data was derived from Mantell with the addition of boldly explicit conclusions that impressed the public and outraged reputable scientists.

Adam Sedgwick had thought the book was so bad that it might have been written by a woman. He was determined to scotch the serpent coils of false philosophy.

> The seductions of the author poison the springs of joyous thought. He has annulled all distinction between physical and moral in the new jargon of degrading materialism. If the book be true, religion is a lie, the labours of sober induction are in vain, human law is a mass of folly, and a base injustice, morality is a moonshine, and our labours for the black people of Africa were works of madness, and man and woman are only better beasts.

Despite the evident lack of knowledge, it caused controversy and upstaged Mantell's *Medals*. The vast torrent of abuse that pursued *Vestiges* was as good as praise *as it sold*. Success came in part because it was lively and energetic and in part because it was so notorious, outwardly athetistic and socially dangerous. Complimentary copies were gratuitously sent to over 200 leading scientific and literary men. Some, such as Silliman, replied on receiving their complimentary copy possibly without having read it with care. Thus Silliman praised it as novel and interesting with many bold conceptions and startling opinions. Others were embarrassed with the opinions on transmutability which appeared too far and too soon. Fanny Kemble, the actress, told Erasmus Darwin that 'its conclusions are utterly revolting to me – nevertheless they may be true'. *Vestiges* was the first thorough-going presentation of the evolutionary theory in English, presupposing that organic creation was progressive through a long space of time. Chambers declared that fossils were not evidence of successive special creations by a God who then controlled all successive creative processes in the universe, but vestiges of creation proving that in aggregate the simplest and most primitive type of life on earth gave birth to the types next above them and that this again produced the next highest and so to the very highest – by which the author transparently intended man.

The work was widely considered to be atheistic and in this respect scientists trod more cautiously. Chambers himself claimed a Deist view supposing the Almighty Deviser had set in place those laws which it is the job of the scientist, and not the theologian, to unearth. Many laymen supported him; thus, to quote the Duke of Manchester in a speech to the House of Lords, 'Are we to suppose the Deity adopting plans which harmonise only with the modes of procedure of the less enlightened of our race?'

Both Owen and Mantell made cautious criticisms. Mantell was humiliated that a work of such flimsy pretensions should have obtained so much consideration, insisting on the dangers of hasty generalisations. 'He is not an original observer; neither is he acquainted with the present state of any one of the sciences upon which he presumes to base his speculations.' Though asked to write a review for the *Quarterly*. Mantell declined as too unwell and too anxious for peace to fish in religion and philosophy for all the errors he would have to expose 'swollen by the upper classes to whom everything boldly asserted and in captivating style is gospel'.

Palaeontologists had struggled to persuade the population that there had been a series of extinctions rather than a single deluge. They had greater difficulty in accepting theories of transmutation, for there was little evidence in the fossil record to support change or increased variation within each stratum. The fossil record particularly for the higher forms of life is very limited. A new species often invade and replace an earlier species, but entire transient populations between new and old are hard to find. Of the fossil population trilobites are complex and abundant and those of the Cambrian period have more primitive characteristics than of the Ordovician and later; but even so it is difficult to catch the appearance of new species in the act of creation. The nearest approach to a demonstration of transition between species has come from examination of the number of lenses in the schizochroal eyes of trilobites.

Towards the end of 1845 a small volume of *Explanations, a Sequel* by the author of *Vestiges of the Natural History of Creation* appeared, in which he asserted that species may look identical during foetal development but in fact were merely analogous. Responding to his critics, he defended his religious beliefs which softened the impact on the scientific community. Mantell even suggested that anyone interested in theories of transmutation should peruse with serious attention and unprejudiced mind the *Explanations* of the anonymous author; but cautioned the reader in making overhasty generalisations on that topic.

A later example of the popular misconceptions about science is shown in Charles Kingsley's *Water Babies* (1863), excited by the controversy of Sir Richard Owen's contention that the great distinctive feature of the human brain was the possession of a structure that used to be called the hippocampus major.

> The Professor had even got up at the British Association and declared that apes had hippopotamus majors in their brains, just as men have. Which was a shocking thing to say; for, if it were so, what would become of the faith, hope and charity of immortal millions? You may think that there are other more important differences between you and an ape, such as being able to speak, and say your prayers and other little matters of that kind; but this is only a child's fancy, my dear. Nothing is to be depended upon but the great hippopotamus test. If you have a hippopotamus major in your brain, you are no ape, though you have four hands, no feet, and were more apish than all the apes of all species. Always remember that the one true, certain, fixed, and all-important difference between you and an ape is that you have a hippopotamus major in your brain and it has none. If a hippopotamus was discovered in an ape's brain, why, it would not be one, you know, but something else.

What had initially been a successful collaboration with the appointment of George Fleming Richardson as curator of his Institute in Brighton was to turn sour. When the collection had been sold to the British Museum, Mantell had helped Richardson in obtaining a post of sub-curator in the mineralogy and geology branch at the British Museum; though at a derisory wage of 7 shillings each working day. After this he alternated between the museum and Brighton where he became a professional lecturer. To quote from Mantell's journal: 'The poor fellow has many good qualities and talents of no common order.' Richardson had given a short course of four popular lectures on geology in Brighton. He wrote well and turned his hand to poetry. Early on he published some of Mantell's observations and inferences in a newspaper as his own ideas and later produced a *Geology for Beginners*, published in New York as well as in England, which usurped much of the material Mantell had collected for his *Medals of Creation*. This aroused Mantell's wrath. His conduct was indeed base; increasingly Mantell received information that he had written to some of Mantell's best friends abusing him most shamefully, and to persons who knew him only through Mantell's introduction. From them, Mantell declared, he has received a severe reprimand but no doubt he had done the same in other quarters. From Silliman it emerged that he had made disparaging remarks about Mantell to American geologists and even in writing to Silliman had attempted to supplant Mantell in his good opinion. Silliman had replied telling him that he had taken an unwarrantable liberty in his assault upon my considered friend.

Although his income increased he was rarely out of debt. Despite his anger tinged with paranoia, Mantell was genuinely and deeply shocked by the death of Richardson at his own hand; found in his bedroom with his throat cut from ear to ear. His suicide was a result of insolvency and the fear of dismissal from his post at the British Museum, a state of affairs which had arisen from a reckless extravagance and the assumption of an unwarranted status.

That Owen's major antagonism was with Mantell probably arose from the fact that until Richard Owen came on the scene Mantell was the only British geologist with an adequate background of comparative anatomy. On the continent Baron Cuvier and his school had studied mammoths and other fossil creatures related to elephants. Agassiz who came to work with Cuvier studied fossil fishes and was later to compare notes with Mantell. Mantell aided the early geologists initially by being able to separate land from sea dwellers, an example being the Cetiosaurus regarded as a whale lizard. He then established that the Iguanodon was an herbivore and was able to piece together the Maidstone Iguanodon and the Hylaeosaurus from their fragmented states. In 1829, Murchison, as President of the Geological Society, asked Mantell's opinion on the fossil fox of Oeningen. In his reply Mantell wrote:

Richard Owen.

> When you first submitted this matchless specimen to my chisel, you expressed your conviction that it would prove to be a species of fox; and as I proceeded in the interesting task of removing the stone from the skeleton, I found myself warranted in agreeing with you in that opinion. When the specimen was entirely exposed, I procured a recent fox and dissected the skull, extremities, etc. and on comparing it with the fossil could detect no essential difference.

He then visited the College of Surgeons to examine various species of Vulpes concluding that it bore a close analogy to Vulpes communis but it may yet belong to an extinct species since specific differences cannot always be detected in the skeleton.

A further involvement in 1851 resulted in antagonism between Lyell and Owen. Mantell's friend, Captain Lambart Brickenden had sent Gideon Mantell a manuscript intended for publication entitled *The Fossil Footprints of Moray*. He had found turtle footprints embedded in the Old Red Devonian sandstone near Elgin and also reported that his brother-in-law, Patrick Duff, author of *Sketches of the Geology of Morayshire* (1841) had discovered similar traces of an extinct reptile in the same strata, six inches long with a curved tail and splayed out limbs which he proposed to call Elginosaurus. Mantell replied that he wished to see more of it to work out its osteological characters and confirm the Devonian strata of the sandstone. 'Could this be borrowed and entrusted to my care? The character of the vertebrae, the structure of the pelvis and cranium, etc. would require an attentive investigation. I will cheerfully undertake this labour.' Patrick Duff undertook to send it down to London to his brother, Dr George Duff, to be seen by Lyell and Mantell. Once it had arrived Mantell wrote back 'though I had but a transient glimpse of it yet I am encouraged to reveal its general character'. After spending three days examining it he wrote yet again. Drawings were made by Lyell and later by Dinkel. 'The reptile is very primitive. I propose naming it Telerpeton elginese from the Greek for remote or most ancient reptile.' He added its ribs are saurian – the pelvis and vertebrae and tail bactrachian. On Wednesday 17 December Mantell exhibited Telerpton at a council meeting of the Geological Society and was to have presented his and Brickenden's paper on it that evening. Dr Duff and Henry Bohn came as guests. Because a very long paper by President Hopkins ran overtime, that on the Elgin reptile was postponed. Owen, who had seen the specimen earlier, carefully inspected it and Mantell's drawings then left the room in a towering rage. Three days later a letter by Owen appeared in the *Literary Gazette*, claiming that Patrick Duff had sent the reptile to London for naming and description by him. Owen called it Leptopleuron lacertinum (slender ribbed reptile). Mantell's letter to the

Literary Gazette was not published but Lyell succeeded in obtaining an apology. To complete matters, at the Geological Society's Council meeting in January 1852 Mantell laid before Hopkins letters from Brickenden and Duff supporting his claim to the specimen.

The most vociferous and vitriolic critic of Owen was Edward Charlesworth who started a new journal the *London Geological Journal* in competition with the Geological Societies' *Quarterly Journal*. In the first edition he assessed the real value and the market price for fossils from Colonel Birch's sale on behalf of Mary Anning in 1820 to that of Mantell's collection to the British Museum in 1838 and others between and after. He juxtaposed Sir Everard Home's perfidy in appropriating unpublished discoveries made by his brother-in-law, John Hunter and burning the Hunterian manuscripts given to his charge; with accusation of mistakes by Richard Owen. He attacked Owen for claiming a Mammalian tooth was from an opossum rather than a monkey; mistaking the skeleton of a modern deer as that of an Eocene ancestor on dental evidence; he used the microscopic studies of J. S. Bowerbank to differentiate the bones of birds from those of reptiles, claiming that Owen had assumed a pterodactyl to be avian and vice versa; and Owen had received a medal for a study on Belemnites which later proved faulty. He added two new chapters on Belemniteuthis from the work of William Cunnington of Devizes and Reginald Neville Mantell's beautiful specimens found at Trowbridge. Despite occasional acknowledgment of Owen's prestige the journal found almost no issues on which Owen had been right.

Owen, as a Hunterian Professor, had catalogued the specimens in the museum of the Royal College of Surgeons; and is best known today for his classification of the saurians, or fossil lizards, into four divisions. Overriding Conybeare's earlier classification, he named the Ichthyosaurs and Plesiosaurs as Enaliosuria with typical lizard-like characteristics such as two openings in the skull. Secondly, he classified the ancient crocodiles, as identified by Cuvier as the Crocodilian Sauria. Thirdly, were the pterdactyls or flying lizards. Finally, as Lacertians or Dinosaurs (terrible lizards), he grouped the Iguanodon, Hylaeosaurus, and Megalosaurus, recognising that their giant thigh bones were quite unlike the curved femurs of the crocodile. The straight, vertical shaft was at right angles to the inwardly turned head which articulated with the pelvis; so that rather than sprawling out sideways at right angles like those of a lizard, the legs rose vertically below the body of the animal. Apart from the name, Dinosaur, Owen was not original. Cuvier had collated Megalosaurus, Iguanodon and Geosaurus but sought further knowledge before classifying them. Hermann von Meyer in his 1832 *Palaeologica* paired the Megalosaurus and Iguanodon as saurians with limbs similar to those of the heavy land mammals. His book was translated by G. R. Richardson, forwarded to

Mantell, and published by Edward Charlesworth in 1837 with Mantell's addition of Hylaeosaurus – a grouping preceding that of Owen.

The story of the Dinornis or Moa, shows both Mantell and Owen in a more favourable light. In 1839 Dr John Rule brought to the Hunterian Museum a six-inch shaft of a femur from New Zealand. Both ends had been broken off. Owen observed the honeycomb matrix and hollow structure and considered that it came from a flightless bird, rather like an ostrich. Four years later a hamper of fossils was sent from New Zealand to Professor Buckland containing bones from a large flightless bird as predicted by Owen. Prince Albert and Sir Robert Peel and society in general were astonished by the height of the Moa (Dinornis). Then in January 1843 Mantell received a box of actual footprints from a Dr Deane together with a long letter explaining their discovery. He presented slabs of the footprints along with the letter to the Geological Society. At the same meeting Lyell reported a letter from William C. Redfield of the United States who had discovered similar footprints in the old red sandstone of New Jersey.

Walter Mantell, Gideon's son, had emigrated to New Zealand and for a while was virtually penniless but settled into government employment. He worked his way to being appointed Commissioner for Crown Lands at Otago on the South Island. Beginning in January 1847, Walter reconnoitred known Moa bone sites in the North Island, living with the Maoris to enlist their help. In September he sent his father a unique and valuable specimen of a moa egg shell which Mantell immediately shared with Owen. Then, just before Christmas 1847, Walter sent his father a box of some 800 fossils; these included the skeleton, a perfect skull, jaws and eggshells of the Dinornis. Furthermore the bones were barely fossilised, suggesting to Mantell that Walter should hunt for a living specimen. Mantell washed and sorted the bones before donating them to Owen who a little more than a week later communicated a paper on the Moa to the Zoological Society. In February 1848 Mantell read a paper on fossil bones from New Zealand before the Geological Society. Mr Charles Darwin was present.

Attitudes to evolution varied within the scientific community; not everyone accepted the views of Lamarck, Saint-Hilaire and Robert Grant when they first appeared. Buckland reasoned that the facts of geology were consistent with those in the biblical story. But others, Hutton, Parkinson, even initially Mantell and Lyell, believed that there had been a series of floods. From this arose Cuvier's doctrine of catastrophes: that successive forms of life resulted from serial episodes of creation and extinction. This remained the belief of Owen and Agassiz. Buffon's *Histoire Naturelle* (1753) challenged Genesis and foreshadowed the theory of evolution, retelling the story of the earth through seven epochs – animals appearing in the fifth and man in the last. Man was an animal who reasoned but was an animal nonetheless; with similarities between apes and man. Alongside the ideas of extinction was another concept that involved species being mutable in time. Such notions developed in France in the eighteenth century; their authors having to combat the wrath of orthodox theologians. There was the disturbing implication that not all plants and animals shared in the grace of Creation. To Buffon the essence of the natural world was constant change, dynamic force and movement. Benoit de Maillet suggested that the underlying homologous similarities of the different structures supposed that they might have changed one into another by specialisation of their parts – wing, hand, flipper – over an immense period of time. (Chambers had used the fact that a giraffe with a long neck has the same number of cervical vertebrae as a pig with almost no neck).

This argument underlay the politics of the Royal Society. Scientific debates were politically important because of the ways in which the world of ideas and the world of public affairs overlapped. The Ultra Tories believed that God exercised a continuous active control over the operations of the world, thus William Kirby in his Bridgewater Treatise on animal Creation postulated a series of 'inter-agents' existing between God and the natural world. Scientists within the establishment believed that their universe was a mechanical one created and set in motion by an external or transcendent God. Other scientists adopted the position that the world had undergone a continuous process of organic development according to the principles inherent in its structure at its creation.

Lastly, some scientists, such as Robert Owen, attempted to adopt a transcendental approach holding at bay Lamarckian theory and that of natural selection. Nature was a great system of progressive development combining creationism, Lamarck's transmutation, animal magnetism, phrenology, etc. This has been called the 'Peelite strategy' partly because of the personal connection, but also because he was willing to make judicious concessions to the Radicals in order to preserve the substance of Establishment thought. It was this pious compromise that the *Vestiges* challenged in 1844.

Among the early popularisers of biological progression or evolution was the unorthodox and eccentric Scottish judge, James Burnett, Lord Monboddo (1714-99) who published works on the origin of language and ancient metaphysics, and advanced the theory of an affinity between monkeys and man. He kept an orang-utan, who figures in his various works as an example of the 'infantile state of our species, who could play the flute but never learnt to speak'; and thereby suggested to Thomas Love Peacock the character of Sir Oran Haut-ton in his *Melincourt* (1817). 'Sir Oran – an orang-outang whom Mr Sylvan Forester, a young philosopher, has educated to everything except speech, and for whom he has bought a baronetcy and a seat in parliament, is an amiable and chivalrous gentleman and plays delightfully on the flute.'

An Oxford don, Dr John Burton, who had difficulty travelling over the notoriously bad Sussex roads, wittily illustrated Lamarck's theories by declaring,

> why is it that the oxen, the swine, the women and all other animals are so long-legged in Sussex. Might the difficulty of pulling the feet out of so much mud have lengthened their bones?

Erasmus Darwin in his *Zoonomia* (1794-96) elaborated on Buffon's ideas. Ostensibly it was a medical treatise aimed at unravelling the theory of diseases, but he also set out his ideas on evolution, believing in a general progress in nature towards greater complexity and perfection. He agreed with Linnaeus that new species arose from existing ones by hybridisation, was aware of selective breeding and believed that there was a constant battle for survival. Linnaeus had revolutionised taxonomy endorsing the idea of species and the crucial notion of a hierarchy of organic life. The animal and plant worlds were linked by shared characteristics, e.g. sexual. Once Erasmus Darwin admitted that there were families of plants and animals he was able to declare that the ape is of the family of man, that it is a generative man, or that man and ape had a common ancestor, like the horse and the ass, that each family, whether animal or plant came from a single stock. Two years before Malthus' *Essay on Population* he declared that Nature was red in tooth and claw – 'one great slaughter house, the warring world' – this was the cost of development. However, in the climate of evangelism Erasmus' evolutionary ideas were looked upon as God-defying fantasies of a crank with cartoons by Gillray showing Erasmus Darwin as an ape carrying a basket on his head labelled Zoonomia or Jacobin plants. Like Lamarck's *Philosophe Zoologique* (1809) he suggested that imprinted patterns of experience were passed on to each new generation, which progressed beyond, in its turn. In the *Temple of Nature*, published after his death in 1802, he wrote of the great length of time since the earth began to exist, perhaps millions of ages before the commencement of the history of mankind.

> Would it be too bold to imagine that all warm blooded animals have arisen from one living filament, with the power to acquire new parts, attended with new properties, directed by irritation, sensations, volition and associations and thus possessing the faculty for continuing to improve by its own inherent activity and of delivering down those improvements by generation to its posterity, world without end?

In *Zoonomia* he had even postulated that 'some birds have acquired harder beaks to crack nuts as the parrot and other birds acquired beaks adapted to break the harder seeds, as sparrows and others for the softer seeds of fruits, or the buds of trees, as the finches'.

Thus Erasmus Darwin foresaw much of evolution and replaced the catastrophic theory by more gradual changes to the earth as described by Lyell in his *Principles of Geology*.

Charles Darwin read his grandfather's book while at university but failed to agree with it until he reread it many years later. Whereas Owen, Murchison and Sedgwick believed that animals could not be arranged in series proceeding from the least to the most perfect, Charles Darwin developed his theory of evolution over many years. On his voyage he had been struck with the observation of the dissimilarity of species of plants and animals inhabiting similar environments. Rather than the simultaneous creation of the fauna and flora of each environment, species could develop in isolation; and frogs, for example, could not cross salt water. Species originated by descent with modification from earlier species; thus well-marked variations gradually turn into the doubtful category of subspecies. Even today the definition of species and how exactly they arise is far from clear. Darwin noted the huge variations within domesticated animals and how different types develop through selective breeding thus developing the hypothesis that feral species gradually evolved over a period of time. He made an entry in his private notebook in March 1838 having witnessed the behaviour of an orang-utang in the London Zoo.

> Let man visit the ourang-outang in domestication ... see its intelligence ... Man in his arrogance thinks himself a great work ... More humble and I believe true to consider him created from animals.

In essence he bore in mind 'how infinitely complex and close-fitting are the mutual relations of all organic beings to each other and to their physical conditions of life'.

Erasmus Darwin had philosophised on evolution well before Charles Darwin's careful compilation of the evidence and ahead of Malthus' *Essay on Population* (1798) based on a struggle for survival. Alfred Russell Wallace in his independent discovery of the theory (1857) also added that ten per cent of man's abilities, e.g. musical talent, do not arise from a clear selective process. Herbert Spencer in a 1851 work postulated survival of the fittest rather than natural selection i.e. the process whereby environmental pressures favoured the survival of individuals displaying particular characteristics; thus setting aside the assumption that a Selector or Creator overlooked the process. Darwin more correctly gave prominence to the struggle for survival and adaptation in the *Origin of Species* (1859) but added that natural selection was the main but not exclusive means of modification. Mendel showed that inheritance depended on genes rather than some temporal averaging of the characteristics of the parents. Weissmann (1886) enlarged our understanding of the germ-plasm; and the twentieth century saw Crick and Watson's double helix of DNA.

Lyell and Darwin both presumed that evolution occurred by gradualism, i.e. small changes occurring through selective variation; whereas evolution mainly occurs through saltation i.e. speciation with new species branching off from a persisting parental stock. Phyletic or phylogenetic transformation when an entire population undergoes change is rarely observed. Our view of evolution has altered considerably since Darwin's day. He and others at the time were too strongly influenced by Malthus and Herbert Spencer into believing that the struggle for existence was paramount. According to Prince Kropotkin (1842-1921) besides the law of mutual struggle there is in nature the law of mutual aid, which, for the progressive evolution of the species, is far more important than the law of mutual contest. This theory of co-existence through varied forms and degrees of symbiosis has been extended even further by Margulis' work on the bacterial origin of mitochondria and by Pain's notable essay on *Darwin's Blind Spot*. We also have a broader understanding of evolution through DNA-genomics.

Sir Charles Lyell's gradualist approach extended to advising Charles Darwin never to get entangled in a controversy, as it rarely did any good and caused a miserable loss of time and temper. As a result Darwin actually wrote in his autobiography: 'I rejoice that I have avoided controversies, and this I owe to Charles Lyell!'

We can ask where Mantell stands in this process. He met Charles Darwin for the first time on 12 May 1837. Lyell, who introduced him, had just published the fifth edition of his *Principles of Geology* which contained a fulsome mention of Mantell's work. Later that evening Mantell lectured on 'The Iguanodon and Other Fossil Remains Discovered in the Strata of Tilgate Forest' to 700 people at the Royal Institution and the next day attended a soirée given by The Duke of Sussex, President of the Royal Society. Their paths would have continued to cross in 1838-9 when Darwin was briefly Secretary of the Geological Society appointed in recognition of his donation of hundreds of specimens to the society. Darwin had yet to publish his *Origin of Species* and, while replete with observations, had not publicly formulated his theory.

LONDON

Mantell had ample reasons to move away from Brighton to the capital. The lease on No. 20 The Steine was due to run out, closing the Scientific Institute and forcing the move of the museum collection to London. The winter of 1836-37 had been particularly severe with snow and avalanches extending into April so that the 'season' in Brighton faded out. He felt very unhappy with 'so much mental misery of late'.

His daughter, Hannah, at school in Dulwich was in a delicate state from February, 'the sea air is too keen for her delicate condition'. At the age of fourteen, while at school, she had developed a long and severe illness with tuberculous disease of the hip joint. Unfortunately, the early symptoms were mistaken for rheumatism leaving Mantell unaware of the severity of the complaint. He found that she had been confined to her room and couch for four months. In April he drove to Dulwich to bring her home.

In May he had lectured at the Royal Society before 700 people on the geology of southeast England, and socialised with the good and the great. He wanted to be part of the scientific establishment and to partake in their deliberations. He watched as the London-based geologists achieved knighthoods and felt he deserved greater recognition, which could only come from attendance at the Societies to which he had been elected and at the various social gatherings. The London-based societies undertook the publication and distribution of tracts and series of volumes informing readers on science. Therefore, Mantell believed he would stand a better chance of getting due recognition of his writings and discoveries in London. The social life of the city also included scientific attractions: some profound, some simply silly. The socialite as well as the scientist (a word not yet in vogue) took an interest in various innovations such as the Thames tunnel, balloon ascents, Wheatstone's telegraphic devices, Herschel's telescope, Babbage's calculating machine, whales stranded in the Thames, and the public dissection of Jonathan Bentham's corpse. To be part of the scene he had to lose his provincial status. Furthermore, with the move of his museum collection to the British Museum, he had a proprietary need to see it safely installed and recognised.

From Brighton he would have learnt exciting news of geological and other events at second hand. For example, Charles Darwin returned to England in early October 1836. Throughout his four year, nine month voyage he had sent back specimens of fauna, flora and geological findings which had been examined by the scientific community. By May 1837 he had finished his *Journal of Researches into the Geology and Natural History of the various Countries visited by HMS Beagle*, based on his extensive notes and detailed diary. It was published two years later. Despite its heavyweight title it proved very popular. He had been feted by Sir Charles Lyell with whom he discussed the evolution of coral reefs and by Sir Richard Owen who had dissected and described the various birds and animals Darwin had found. He conceived the idea of a multi-volume work with 150 coloured plates and the Chancellor of the Exchequer gave him an open grant of £1,000 enabling him to publish his zoological and geological findings (much more forthcoming than the delayed price to be paid for Mantell's enormous collection given to the British Museum). 'How contemptible,' he thought, 'is that legislature, which with the enormous riches of this country, cannot afford the petty sum to obtain what is peculiarly British and unique.'

However, the chief cause for his move was the lack of success of his medical practice and his inability to make a satisfactory living in Brighton. He made his first public announcement of his intended departure in June when conducting an excursion with the Sussex Royal Institution to Shoreham, Bramber and Steyning, describing the interesting old churches and archaeological details. Quoting from Childe Harold, he declaimed:

> Farewell, a word that must be and hath been
> A sound that makes us linger, yet Farewell!

In September 1837 he finalised plans to purchase a practice at Clapham Common, three and a half miles from Westminster Bridge, after considerable wrangling. As he explained to Silliman, the sum required was generally equivalent to two years' profits. He paid £1,500 to Sir William Pearson, surgeon, apothecary and accoucheur, who had decided to retire. £1,000 would be paid within three months and £500 with interest a year later. By contrast, Mantell eventually sold his Brighton practice for a mere £105. Sir William had been in practice for fourteen years and was to remain till the summer of 1838 in order to introduce Mantell to his patients before

Clapham Common practice.

retiring in his favour. Sir William Pearson may have helped initially but by December 1840, Mantell was writing: 'That mercenary creature Sir W. Pearson attempted to interfere with my practice, and compelled me to have recourse to legal proceedings to protect myself from his contemptible conduct.'

To the original draft Pearson had added a provision enabling him to resume practice at Clapham after two years; unsurprisingly this had been refused and both sides argued through attorneys for several months before agreement was reached in January 1838. A rider was added to the contract establishing a means of arbitration in case violation of the agreement was alleged by either party.

In August 1841, he had still more legal troubles – an action for an alleged debt was brought against him by a 'vagabond fellow', a druggist, which was decided in the Secondary Court in Mantell's favour. The vexatious affair had been going on ever since February, and he had been greatly annoyed and put to a great expense. 'Yet the man had not a shadow of a pretence to claim anything from me, and on the contrary he is indebted to me about forty shillings. Yet such is the state of the law and the villainy of the lawyers.'

As early as 1834, he had contemplated a popular book on *The Wonders of Geology*, subtitled *A Familiar Exposition of Geological Phenomena*, based upon his lectures in Brighton and inscribed, hopefully to the Earl of Munster, son-in-law to Lord Egremont – a little popular work, the romance of Science; and also his farewell to geology. Although he later declared that he had written every word, he put forward G. F. Richardson's name as editor, perhaps to suggest that he had dictated it while concentrating on his medical practice. It was written in a frankly popular style with mass appeal, while hoping to capture some of the beautiful language of the poet Felicia Hemans (1793-1835), remembered for the sublime opening lines of her poem 'Casablanca' – 'The boy stood on the burning deck / When all but he had fled'. It was narrated as from an imaginary cosmic higher intelligence looking down on Brighton and Southern England from a flying carpet and viewing successive mutations of the landscape from the period when the Portland forests were flourishing.

The first edition of a thousand two-volume copies, published by Relfe and Fletcher, appeared in April 1838, the second thousand in May. All in all there were eight editions, containing various modifications, the last being published after his death. The sale of his work exceeded that of all the published oeuvres by Bakewell, Lyell and Phillips during the same period. There was an American edition and a German translation. The frontispiece was an engraving by John Martin of the Country of the Iguanodon. Silliman considered that:

> in point of science it is precise, accurate, condensed and cumulative in proof, no important facts are omitted and none re unduly expanded. However it was still only a popularisation, furthering public understanding rather than knowledge itself.

In the first lecture, he describes the theories of the formation of the solar system as stated by Herschel and Laplace; agreeing with Lyell that planet Earth, and the universe, were of infinitely greater antiquity than any literal reading of the Bible might suggest; though he still considered it necessary to discuss the harmony between Revelations and Geology and to introduce a passage from a sermon by the Bishop of London. He discussed the earth with its geological epochs while noting that in our own time the Dodo and the Irish Elk had become extinct.

The second lecture discussed the fossil bones of large mammals, such as the mammoth, mastodon, Megatherium and Dinotherium. The third considered the tertiary finds of the Eocene, Miocene and Pliocene with marine and freshwater shells. Next came the age of reptiles with glowing descriptions of the country with its Fauna and Flora. In the fifth lecture the rocks of the new red sandstone, carboniferous measures and limestone were shown to yield crustacean, trilobites, crinoids and corals. The final lecture began with further remarks on the strata of the Old Red Sandstone and the older rocks as seen at Fingal's Cave and the Giants' Causeway, and ended with earthquakes, volcanoes and the production of ores and gems.

His daughter, Ellen Maria, had done most of the drawings for the *Wonders of Geology* despite being in delicate health and staying with Dr George Mantell, a cousin of her father, who practised in Farringdon in Berkshire. The invalid Hannah was still suffering, disabled by disease of the hip and wholly confined to bed. Reginald was at school in Clapham. Mantell's health and that of his wife were better than they had been and he saw no reason why he could not accomplish more and better things.

In 1840 he was 'somewhat mortified' when the President of the Geological Society in his annual address omitted all reference to his book although it contained, even in the drawings alone, more important discoveries in fossil remains than any other book of the season[1]. Later in the year he was introduced to Prince Albert and presented him with the German and English editions of the *Wonders*. Mantell commented that 'he is a very affable, intelligent, handsome youth, and those who are intimately acquainted with him assure me he is very amiable; what a pity he should be exposed to such as Court as ours!'

Although Mantell was able to profit from the practice after April 1838, he had to sell his collection only to be told that the £4,000 would not be forthcoming from the government for a further year. In addition the £150 profit from the fifth edition of his *Wonders of Geology* was delayed for 18 months.

He left Brighton on 12 March taking his younger son, Reginald, with him. His wife and daughters were to follow at the end of the month but Hannah was still very ill and it was doubtful whether she could make the move. He felt his health had improved, enabling him to pursue his scientific and other interests. Viewing the extensive practice he declared 'that the neighbourhood is so populous, wealthy and respectable. I have met many friends and indeed the reception has been very flattering'. His scientific background appeared to be an advantage and he held two soirées, gossiping on scientific subjects. Once settled he hoped to resume his geological researches, for he had become such a favourite that he had nothing to fear on the score of his profession.

He was delighted with his new house, Crescent Lodge, very agreeably situated on the main road from Brighton to London, which he had taken on a five year lease. Installed in his study he was surrounded by his old bookcase, his first glass case for his Tilgate fossils, a bust of Lord Egremont and the walls held diagrams for his lectures and skeletons of fishes and ophidians (snakes). The chimney piece contained Norman tiles from Lewes Priory.

His practice gradually improved with the addition of new clients, lifting him somewhat out of his depression. In September he wrote:

> My professional prospects are better; and although not so good as they ought to before the tremendous sacrifice I have made, yet they are so favourable as to leave but little apprehension for the future.

Later, when in a more elated mood, he declared that he had been fully and profitably engaged in his profession, with his usual good fortune that is, in regard to the successful treatment of his patients. He had had many and important cases and had been attending with Sir Astley Cooper 'and other of our eminent surgeons' and had gone through the ordeal triumphantly having succeeded in many difficult operations and been highly successful.

However, he was sadly broken up in health and energy. The labour, which a few short years ago would have been mere amusement, is now most fatiguing. Elsewhere he says that he has been almost dead from fatigue, anxiety and that sorrow which language cannot describe. Some months later he wrote:

> I have in some measure recovered my tranquillity of mind but alas I have no home for my affections. I have acquired several kind and most valuable friends here, and am highly respected.

He had had severe fainting fits, had suffered excruciatingly from tic douloureaux and all the attendants of a highly excited state of the nervous system but was now better. Though when Mr George Kingsley, son of a colleague of Silliman – 'an intelligent and interesting young man, by profession a lawyer established at Cleveland Ohio near Lake Erie' – visited London on business and requested that he might be introduced to Mrs Mantell and the rest of the family, Mantell said that he was greatly disappointed that he could not meet Mr Kingsley as he 'was so ill last week on the day he wrote me that I was obliged to decline his visit'.

While it is probable that Mantell was psychologically, and possibly physically, ill and deeply sunk in misfortune, the real reason was unearthed by Dean while examining Mantell's copy of William Berry's *County Genealogies of Sussex* (1830). Though erased both from his letters to Silliman and from his journal, there is a line-and-a-half entry partially deleted that records the date when his wife, Mary Ann, after twenty-three years of often stormy marriage and several extended separations, walked out forever on 4 March 1839, together with a despairing note on the loss of his museum:

> There was a time when my poor wife felt deep interest in my pursuits and was proud of my success but in later years that feeling has passed away and she was annoyed rather than gratified by my devotion to science!

In a letter to Silliman of the same date he stated that he was overwhelmed with professional engagements, including a number of important cases. He was frequently working with Sir Astley Cooper and other prominent surgeons and been highly successful. However, he was now 'almost dead from fatigue, anxiety, and that sorrow which language cannot describe.' He had also experienced severe fainting spells, excruciating facial neuralgia, and other symptoms. His housekeeper, Hannah Brooks had left with Mary-Ann to live in a small cottage near Exeter. The household was in turmoil. Hannah Matilda was sick, Reginald attended school nearby, Ellen Maria returned home after the end of the spring term at Dulwich but left in September to live with Mrs Richardson in Dulwich and resume her schooling. Walter's apprenticeship in Chichester was ending but he refused to enter into partnership with his father and on 15 September sailed for New Zealand as a pioneering immigrant, never to return. It was Mantell's intention, once Walter had completed his medical studies, to indulge him in a trip to America 'ere he settled down to practice here as I hoped'. Added to this, his brother, Joshua, who had been under Thomas' care for thirty months, finally required admission to an asylum at Ticehurst, Sussex in

April of that year. Mantell was required to make annual payments of £54-60 for his care as a patient.

Mantell's distress is given added credence by his lack of diary entries over that time and his friends became aware of his predicament. Later that year Bakewell wrote to Silliman: 'I am anxious about Dr Mantell, I have not seen or heard of him since his son sailed for New Zealand in September'. Lyell tried to assure him that the prospects for Walter in New Zealand were most auspicious.

Hannah Matilda was confined to her bed unable to walk, obliged to lie on her back but able to draw, paint, knit and write. During the summer she appeared to improve and her disposition remained calm and sweet. He was pleased with her progress and writing to Silliman he pondered:

> How mysterious are the ways of Providence! If there was ever a human being free from the waywardness of temper and the usually failings of mortals it is that sweet girl! Is it to teach us that by patient suffering we can alone be made perfect?

He constructed an invalid carriage to enable her to go out in the open air. He would bathe and cleanse her wound for an hour each morning and evening. That autumn the hip became so wasted with infection that the bone was painfully prominent through the skin. He moved his bedroom next to hers and rarely left the house except when called away professionally.

In March 1840 he dined at the Athenaeum, London's literary club, to which he had been elected the previous month 'under the most flattering circumstances'. But he was anxious about his daughter so returned home early, and within ten minutes was summoned to her side as she had fainted with a sudden and profuse haemorrhage from the abscess about her hip. She awoke the next morning and appeared to enjoy her breakfast but was very weak and in imminent danger. Mantell's sister and niece never left her, but one morning she felt faint again and whispered her father's name wishing him to come from his bedroom. He ran into the room just in time to hear his daughter's final sigh. 'Her gentle spirit passed away'. For months he remained in a state of depression, frequently visiting his daughter's grave. At fifty, he appeared careworn and it took many months to recover some measure of tranquillity of mind. 'I have no companion; no one whose smile or approbation would cheer me on.'

It is worth repeating Spokes' comments as they provide an insight into Gideon Mantell.

> A good deal of what happened at Crescent Lodge in 1838-9 must go unrecorded. We are dealing with Gideon Mantell as a Surgeon and a Geologist and personal matters of health and fortune are only introduced to show the difficulties under which he struggled while carrying on his professional work, at the same time as his scientific researches and writings. It is therefore necessary only to allude to the fact that domestic occurrences of the most serious kind must have had a corresponding effect upon a man constituted as Mantell was. In his anxiety to succeed in his geological pursuits, his domineering character was impressed upon the other members of the household to an extent which ultimately may well have caused resentment.

In a letter to Silliman dated 2 September 1840 he explains his reactions:

> My health is somewhat improved ... Last week the first and only intelligence of Walter since he left England dated April 6th ... I have been from my youth up a water drinker and very temperate in my diet. Of late I have allowed myself two glasses of light wine in water daily, but that is all I take, and I am certain it is by far the best method. As we advance in life stimulants are necessary. But he who is accustomed to them in early or middle life, requires with advancing years more than the constitution will bear, as the termination of the state of being instead of being 'holy calm in the night' is a scene of protracted suffering.

Later in the year, Mantell's fundamental tenacity of spirit reasserted itself. He went with Reginald to the Isle of Wight, visiting Alun Bay, Shanklin Chine and Ryde and collected fossils. In London he was able to participate regularly in meetings of the Geological Society and agreed to lecture on scientific subjects in Clapham. After one such lecture on the 'Interest and Importance of Geological Researches' he was gratified to receive the support of the local church and agreed to give a further lecture at the Clapham Boys Parochial School. He gave other lectures in the splendid picture gallery of the house of a wine and brandy merchant John Allnutt, gaining the friendship of Mr Allnutt and his wife. Another friend was Catherine Foster a subscriber to the South Downs who conducted a seminary for young people and was the sister of Reginald's schoolmaster.

In November Mantell assisted Owen in his endeavours to collect material for a comprehensive report on the *Fossil Reptiles of Great Britain*. Improved communications with rail links to all the major cities meant that Owen was able to examine many private fossil collections and gained considerable kudos by correcting egregious mistakes by his predecessors with regard to the comparative anatomy of the various fossils. However, while performing this task, Owen all too often failed to take into account the work of others from whom he sought help; as was manifest in his classification of dinosaurs.

For a time relations between Mantell and Owen were so cordial that Owen invited Mantell to dinner with William Buckland. During the evening Owen showed his envious guests his new microscope and they examined each other's blood. Mantell's erythrocytes appeared almost macrocytic (i.e. larger than the others). Dr Buckland exclaimed, 'Why Mantell you see you have a good deal of the reptile about you!'

The microscope was the new research tool of Science. Galileo had used two lenses each ground in a double convex shape and recognised that if properly arranged they could magnify tiny objects. He called the instrument an occhiolino (little eye) but his friends suggested the word microscope. Drawings taken by Galileo with his primitive microscope were published in Rome in 1625.

Two names were particularly associated with its application. Anton van Leeuwenhoek (1632-1723) is described as the father of microscopy. He was a draper's apprentice using magnifying glasses to count the threads in cloth and taught himself to polish and grind tiny convex lenses the size of a pin head with a lot of curvature. These were fixed between two metal plates with suitably placed apertures. A system of screws allowed the user to move the plates and focus the specimen. The apparatus was held close to the eye giving a magnification up to 270 times. Robert Hooke (1635-1703) published a book entitled *Micrographia* in which he coined the term 'cell', likening plant cells to the confined discrete cells of monks. He conceptualised light as a very short vibrative motion transverse to the straight lines of propagation through a homogenous medium and accused Newton of plagiarising his ideas of the rainbow. The first microscopes in the seventeenth century were tall with separated lenses. His design for a microscope was utilised by van Leeuwenhoek who in turn described bacteria, yeasts, the teeming life in drops of water, and the circulation of blood corpuscles in capillaries, reporting his findings to the English Royal Society and the French Academy. The next century saw further refinements. Joseph Jackson Lister in 1830 developed achromatic lenses, reducing spherical aberration by using several weak lenses together at certain distances without blurring the image. Mantell records how, on 27 August 1833, he called with his daughter Ellen on Mr Joseph Lister who was staying at Brunswick Terrace in Brighton. Mr Lister had been investigating the circulation of flustra and other zoophytes with his beautiful microscope, and they had the great treat of seeing several beautiful drawings of various zoophytes, and moreover, in the animals themselves which Mr Lister's instrument displayed in a more clear and admirable manner than he had ever before witnessed. In Holland, Harmanus van Deyl fabricated lenses combining two types of glass, reducing the chromatic effect and the disturbing halos resulting from differences in the refraction of light. From 1833 it was possible to use the oxy-hydrogen microscope for public display and throughout the 1830s and 1840s its use proliferated in institutional lectures. The illuminating power was a powerful and steady lime light formed by a ball of lime heated to whiteness by a flame of hydrogen and intensified by a jet of oxygen.

In his correspondence to Silliman, Mantell gave an amusing account of microscopic investigations which had become the fashionable craze.

> At the Royal Society there was sure to be a paper on the microscopical investigations of the embryo, at the Medico-Surgical Society microscopical observations on the blood corpuscles, at the Geological microscopical observations on fossil teeth and at the Linnaean microscopical observations on vegetative organisms! At the soirée given by the President of the Royal Society (the Marquess of Northampton) microscopes were the great attention and there was a Microscopical Society and a Microscopical Journal which by the bye is edited by a man who cannot write grammatically.

Despite all his bluster and bluff, Mantell quickly became an addict and introduced Reginald to the microscope. He had been examining some water from West Point in America sent by Dr Baily to a friend of Mantell's in London.

> We have seen some of your Infusoria alive and as active as if they had not had a sea voyage and been bottled up for weeks. If you know and are near Dr Baily some of the said water with Zanthidia would be a great treat to me. Prince Albert was at Lord Northampton's soirée and I showed him many fossil teeth and bones and infusoria with the microscope. He is a very pleasing, intelligent young man, exceedingly affable, very handsome and elegant. He should consider that a short course of the instruction in the use of the microscope might act as an antidote to the "coursing, hunting and racing" which occupied so much of the time of the young Prince Consort.

Owen examined minute slivers of Iguanodon teeth comparing them with those of a modern iguana and showing a very different internal structure. Similarly, when comparing samples of forearm bones with those of modern reptiles, the Iguanodon bone was more analogous to those of herbivorous mammals. Mantell's interest in the large fossil reptiles had not diminished despite his obsession with the microscope. In the *Transactions of the Royal Society* of January 1841 he reconstructed the environment and mode of behaviour of the Iguanodon having collected the remains of some seventy specimens. He believed that it stood upright on enormous hind legs using its slender, prehensile forelimbs to gather the foliate from the tall trees and ferns. To a friend he surmised that it had a long acquisitive tongue and grasping lips. He augmented his original discovery of Hylaeosaurus with two further major finds and examined several fossil turtles; and had the satisfaction of being elected to the Council of the Geological Society.

*

Owen examined not only Mantell's collection at the British Museum but Saull's in London, Holmes in Cuckfield and countless others in different parts of the country made possible by rail travel. His report to the British Association for the Advancement of Science (BAAS) on British fossil reptiles was given in two parts. In a two and half hour presentation in Birmingham in 1839 he concentrated on the anatomy of sea lizards, originally classified by Conybeare as Enaliosauria (marine lizards), and identified ten species of Ichthyosaurs and sixteen plesiosaurs. He was warmly congratulated as the greatest comparative anatomist living and granted £200 by the BAAS.

In a similar lengthy presentation in August 1841 to the BAAS at Plymouth he discussed amphibious and terrestrial sauria as described by Buckland and Mantell dividing them into Crocodilians, Dinosaurian, Lacertian, Pterodactylian, Chelonia (fossil turtles), Ophidian, and Batrachian reptiles, superseding the classifications of Cuvier and Hermann von Meyer. He renamed the fossil crocodile, Suchosaurus, adding Goniopholis and Poekilopleuron from Mantell's collection. Both Mantell and Owen modified their conception of the Iguanodon, once thought to be 100 feet long. The shaft of the femur fitted into the pelvis at right angles giving a vertical stance, not sprawling sideways as with crocodiles, and the sacrum was fused enabling easier weight bearing. Mantell's bipedal stance was not widely accepted. Owen's distinct group of dinosaurs (terrible lizards) – the term dinosaur first appeared with the publication of the report and was not used at the BAAS – consisted only of Iguanodon, Hylaeosaurus and Megalosaurus, omitting among European dinosaurs Streptospondylis, Cetiosaurus, Thecodontosaurus and Poekilopleuron. The Association received the report with raptures and granted him £250 for engravings and a further £250 for a further report. Mantell, who had the misfortune of his daughter's death and the break up of the family, as well as the need to develop his medical practice, was unable to attend either meeting.

Although Owen acknowledged him in his presentation, in redrafting the report for publication he carefully rewrote Mantell out of consideration citing the Iguanodon as found by Cuvier. The publicity Owen gained from these reports made him a star of Victorian England, idolised by the public and known even to Royalty.

The beginning of 1841 revived Mantell's demeanour. In May at a soirée at the Allnutts' with 100 people present he chatted for an hour on Roman vases, incrustations, fossil turtles, Infusoria and the use of the microscope. He visited the opera house and spent a lovely day at the Horticultural Gardens at Chiswick accompanied by Reginald and meeting Samuel Rogers the poet, noted for his *Pleasures of Memory* and Italian verse tales; Charles Babbage the mathematician, who had developed his calculating analytical engine; the Bishop of Exeter and the Duke of Cambridge. He attended the Royal Society where his paper on fossil turtles was read (or rather mumbled) by the chairman; called on Lyell on the eve of his departure for America, and visited Mrs Murchison who had just received a letter from her husband giving an account of his reception at the court of St Petersburgh on his way to the Ural mountains.

Private receptions in the mid-nineteenth century in London and the provinces were frequently termed soirées or conversazioni. Both refer to an evening gathering of specially invited guests entertained by conversation, various displays and where social introductions were made in profusion. Many of the eminent guests were encouraged to bring artefacts, antiquities, or scientific apparatus and findings which they could then discuss. The displays and artefacts would be laid out on tables or benches so that the throng of people could circulate, chat, view and discuss all that was on show. Musical soirées were separate occasions but there might be a small orchestra playing on arrival. Food and drinks were usually available and the timing could be strictly adhered to. If there was a difference between the two terms, conversazione might be a more intimate function and soirée a larger get-together. Much of the success and social distinction of the Royal Institution as a high level centre of popular science stemmed not from its courses of lectures, but from its Friday evening conversazioni begun in 1825, attracting new subscribers to lift the institution out of longstanding debt. Likewise, the London Institution attempted to duplicate that of the Royal Institution with a Friday evening programme from 1828. In the conversazioni a wide range of topics was discussed, concentrating on new, easily illustratable technology or biological subjects.

In June 1841, Mantell travelled by rail for the first time taking Reginald, leaving from Paddington for Chippenham, Bath and Bristol and returning to Farringdon. In July he made the first experimental trip to Brighton with the opening of the new line; and on the 19th went to Swindon to ramble alone and unmolested over the scenes of his early childhood. He set off on a walk to the churchyard and along the pathway leading to the long walk, the scene of many happy hours with his first love, now long since numbered with the departed. Afterwards, he went to the quarries and worked there till 4.00 p.m. and at six, walked to the railway station with a heavy load of fossils which one of the quarrymen carried, reaching London at nine, taking tea at the Athenaeum and returned home by eleven. In a similar attempt at fossil gathering in September, he went by the Brighton railway as far as Haywards Heath walking across the fields to Whiteman's Green on a fine hot day but no men were at work and there was not a fossil to be found.

Mantell vacillated in his attitude to the railway system though by the late 1840s he was living a life based on express timetables. By the end of the decade he would take the ten o'clock from London to Brighton to deliver a lecture at the Town Hall at 3.00 p.m. complete with drawings and specimens and would return home by half past nine the same day. On the next occasion he would leave London Bridge Station reaching Brighton by 2.00 p.m. With travel so much easier the significant architecture of the period took the form of Railway termini and hotels. He wrote of that 'splendid railway the Great Western by which geologists may be transported in five or six hours from the Tertiary strata of the metropolis to the magnificent cliffs of mountain limestone at Clifton.' The additional fact that his son was employed developing the system should have led him to be enthusiastic. But he could also write, as in 1845 of express trains, terrific to behold, raising clouds of dust from the whirlwinds induced by their rapid passage through the air. His antipathy, overruled in practice, arose partly because he felt it was diverting money which should have gone to alleviate famine in Ireland and fear at over-speculation possibly leading to a repeat of the South Sea Bubble. 'Everyone is mad upon the subject, thousands are speculating on them and many have made fortunes within a few days.'

On 10 October, Mantell drove through a rain storm to Craven Cottage, Fulham to dine with Sir Edward Bulwer-Lytton with whom he had corresponded.

> The cottage is charmingly situated on the bank of the Thames, and is very recherché. A central vaulted apartment, called the Egyptian Hall fitted up imaginatively a-la-Egypte, with a drawing room on one side and a study-library at the rear, all richly furnished in admirable taste, and containing many choice objects of art and vertu.

Sir Edward received him with great cordiality. Mantell described him as about middle height, elegant in person and manners, spare in person, with dark hair and whiskers, an acquisitive nose, finely chiselled smooth chin and evidently of delicate physical organisation. With just three others present they conversed that evening on a great variety of topics, including animal magnetism, geology, Shakespeare's plagiarisms, the microscope, the telescope, literature and drama. He left at ten o'clock much gratified with his visit to one whose writings had so often delighted him and beguiled the hours of sorrow and sickness.

Catastrophe struck the next day, the details of which will be analysed carefully. His journal entry for 11 October merely states:

> Narrowly escaped severe injury; from a fall occasioned by jumping out of my carriage to extricate my horse, the coachman having allowed the reins to become entangled. I fell with great violence, the wheels just grazed my head but did not pass over me.

Not feeling seriously injured, he continued with his duties, attending a lady with concussion, and afterwards walked home on an intensely cold night. Pains already in his lower back intensified, and his left foot became numb, with an ascending paralysis. By the 28th he wrote:

> The paralytic symptoms are more severe, the numbness and loss of feeling extended from the feet. I can scarcely walk. The pelvic viscera are dreadfully affected.

Except to seek treatment with hot baths and brisk purgatives, he stayed at home. In the succeeding weeks the paralysis and numbness came and went, sometimes extending to his right arm. The lack of feeling spread to the gluteal muscles and down the insides of the thighs and leg. And by November the symptoms were accompanied by severe spasms and cramps. By 5 November there was an improvement in the numbness with more sensation in the limbs and pelvic viscera.

He explained his predicament in a letter to Silliman on 17 November:

> For the last three months I have been suffering from a paralytic affection of the lower part of the body; the effects have been most distressing as you will readily conceive. For several days I could not walk without support. By the use of hot baths and suitable remedies, I am now, thank God, very much better, can walk steadily and although very weak, and suffering from numbness of the limbs and spasm, may be considered convalescent.
>
> But I cannot doubt that this attacks is connected with an affection of the spine, the result of my laborious attention to my poor child and to

> some other distressing cases at the same period, and from the effects of which I shall never recover... my sister... my servants, and my assistants have all shown every attention, and several kind friends have afforded me much sympathy, and supplied me with many little luxuries, enhanced by the kind spirit which dictated the act. I am so thrown into confusion by this horrible attack that I cannot resolve upon anything. I have sent for Reginald who is now with me.

Silliman was as for ever supportive. In his reply later in the year, he asks:

> Can you not place some professional friend as a locum tenens in you place while you embark in the spring for this country and pass a few months or weeks, if you can spare no more time with us, making our house your home and regulating your excursions as you may find it convenient and agreeable, and if Reginald or your daughter should accompany you, so much the better. As to Owen's behaviour: his treatment of you is unjust, dishonourable and merits exposure.

Even when his mobility improved he remained an invalid. He could not stoop, or use any exertion without producing loss of sensation and power in his limbs. The symptoms did not abate and he felt almost dead from pain and fatigue and 'a tumour of considerable size' began to develop on the left side of his lumbar spine. When the Lyells saw him after their trip to America they were surprised by his ghastly appearance. Feeling thoroughly incapacitated on 7 December 1841 he forced himself to deliver a promised lecture on corals and animalcules at the Parochial School in Clapham. At his second lecture that month, 260 people attended. During this period he was mostly confined to his house but on 12 March 1842 felt well enough to enjoy a soirée given by the President of the Geological Society, Roderick Murchison, in Belgrave Square; and he lectured four days later at the London philosophical Institution. Although he attempted to get about, there were days when he felt very ill, scarcely able to visit his patients, obliged to give up attendance on most of the scientific meetings, and others, such as 4 May 1842 when he was confined to his bed – 'Very ill with neuralgia of the heart; took seventy-five drops of laudanum and twelve drops of prussic acid before any relief was obtained.'

He explained his neuralgic pains and paralysis in a letter of 18 October 1842:

> From the pressure of the gathering on the roots of the motor nerves, the paralytic symptoms are occasioned; by the irritation of the roots of the

> sensitive nerves the dreadful attacks of neuralgia from which I suffer daily and almost nightly so as to deprive me of sleep for a week together. If I could give up all professional visits, much of my suffering would be avoided, but this I cannot afford to do...So much of the frail tenement in which my soul is imprisoned.

In addition to the neuralgic pains and spasms, the result of compression of nerves to the lower limbs as they came out of the exit foramina of the spinal column, on September 18th 1842 he complained that

> while still suffering a tumour of considerable size has gradually made its appearance on the left side of the spine, it now evidently contains fluid – probably a lumbar abscess; and my long probation of suffering will be terminated by a painful and lingering death.

Four days later Robert Liston FRCS (1794-1847) (known for his splint) and Sir William Lawrence (1783-1867) President of the Royal College of Surgeons examined his back declaring there is a lumbar abscess but one of them considered that it might not be connected with diseased bone. Remaining still very ill and languid and scarcely able to get about, he consulted Sir Benjamin Brodie on the 28th only to get a further conflicting opinion.

> Well may the un-medical sufferer exclaim "Who shall decide when doctors disagree" – when I am at a loss to decide upon the conflicting opinions upon my case Sir B. Brodie and Mr Lawrence believe there has been periostitis and there is diseased bone, but not of the bodies of the vertebrae and that an abscess has formed and will probably require to be opened and perhaps some exfoliation will take place. But the one recommends entire rest in a horizontal position and blisters moxa on some external stabilisation while the other advises no external application, gentle carriage exercise and sarsaparilla.
>
> Liston and Coulson (1803-77, Senior Surgeon at St Mary's) think there has been no disease of bone but simply the formation of an abscess which will be discharged and get well. I have no doubt of inflammation of the theca spinalis arising from great fatigue and stooping over Mr Hebbert and my poor daughter was the origin of my suffering; Lymph or serum was poured out, and at length pressed on the origin of some of the lumbar nerves, and gave rise to the paralysis; that pressure is lessened from the infiltration of the effused fluid under the great lumbar muscles and the paralytic symptoms are relieved but that there is also diseased bone, probably on the outer ring of the diseased bodies of the vertebrae

> and a collection of pus; the dreadful attacks of neuralgia arising from irritation of the sensory nerves of the pelvis and lower limbs from pressure of diseased bone or confined effused fluid – The perfect rest I ought to have cannot be obtained.

To Silliman he again quoted Mrs Hemans:

> Thou art like night, oh sickness! deeply stilling
> Within my heart the World's disturbing sound
> And the dim silence of my chamber filling
> With low sweet voices, in life's tumult drowned!

The following year on 14 August he once again consulted

> Sir B Brodie and Dr Hodgkin. Brodie found the tumour larger, but nothing could be done but rest (alas how can I obtain it?) Dr Hodgkin thinks I shall recover and advised iodine. Mr Lawrence was very kind and recommended as little fatigue as possible and not to open it until the skin becomes thin.

On 29 April he went to Mr Lawrence with his cousin, Dr George Mantell. They found the tumour larger and decidedly fluctuant so that a lumbar abscess is inevitable. In May 1849 he again consulted Mr Lawrence who examined the tumour and pronounced it much worse, advising that he ought to give up all professional exertion.

In 1842 the gratitude of the people of Clapham Common for his lectures was shown with the presentation to Dr Mantell of a Ross microscope worth ninety guineas by a local barrister. The idea originated from Rosina Zornlin, author of Recreations in Geology (1841) together with her sister, the Allnutt family and Miss Foster. As recorded in *The Lancet*:

> Testimonial to Dr Mantell. The clergy and principal inhabitants of Clapham Common and its vicinity have testified their sense of the advantages derived from Dr Mantell's lectures on Physiology and other branches of science, by very appropriately presenting to that gentleman, who is one of the most persevering and successful cultivators of science in this country through the Rector, the Rev. Dr Dealtry, a microscope and apparatus value 100 guineas; with the following inscription. "Presented to Gideon Algernon Mantell Esq., LLD, FRS &c, by his friends in Clapham and its vicinity in testimony of their grateful sense of his kind and effective exertions among them for the advancement of scientific knowledge".

Among the subjects he had lectured on to the people of Clapham Common were coral reefs and foraminifera. Coral reefs were a favoured topic of Charles Lyell though he had to modify their origins in the light of Darwin's observations. It is now known that the reef-building corals are exclusively symbiotic with unicellular yellow-brown photosynthetic organisms known as dionmastigotes which live within the coral's gut. Foraminifera could be described as free swimming corals forming a group of aquatic life forms with over 35,000 species, a few forms are found in freshwater but are more widely spread throughout the oceans and sharing a similar relationship with photosynthetic organisms. They are amoeba-like protozoa. The protoplasm is not differentiated into an inner and outer layer and some have a nucleus. They feed and move by means of protoplasmic processes known as pseudopodia extruded through the shell openings. The minute external shells of remarkable beauty and complexity are no bigger than grains of sand, and typically calcareous and chambered with small holes or foramina. In their billions they form an important component of plankton, the basic food source of marine life. The marine remains which include many fossils form a sediment on the sea bed, known as foraminiferal ooze and are involved in the deposition of limestone.

SOIRÉE

Mantell gave three talks on topics which currently interested him, at an extraordinary soirée held by the Allnutts two days before Christmas 1842. On 1 July he had received, from Sir Woodbine Parish[1] of Naples, a fragment of the Column of the Temple of Jupiter Serapis at Puzzuoli near Naples. He recorded that he had obtained

> a fine perforated fragment of a column from the Temple of Serapis and am expecting daily a model of the Temple made on the spot by an Italian artist. I think I never had a specimen that delighted me more. It is in a glass case before me now, and all the fine associations connected with its history crowd upon me.

He was also delighted on 13 July to receive two letters from 'my runaway son Walter', and from Mr Charles Dickens (Boz). Unfortunately he does not tell us what it contained.

Lyell had made the Temple of Serapis his frontispiece as an example of geological causes in operation at present, which he then used to explain phenomena of all ages. The Temple had been built on land, sunk into the sea and pierced by marine molluscs then once again uplifted and exposed within historic times. The model and a sketch made by Mrs Allnutt on the spot were displayed. It also appealed to Mantell's antiquarian interest. Serapis represented two Egyptian deities, Apis and Osiris, and was further compounded by features of Greek gods such as Zeus and Dionysus. The gods were introduced to Alexandria by Ptolemy in an attempt to unite Greeks and Egyptians 'in one common worship'. At Puzzuoli, the Roman supreme god, Jupiter, was added.

His second topic, fossil infusoria, combined his interests in palaeontology and microscopy. The President of the Microscopic Society, which Mantell had joined, the Rev. Joseph Reade, had visited him that afternoon. Infusoria are protozoa and rotifers found within organic material. This followed Buckland's study of coprolites and pseudo-coprolites, supposedly the droppings of Mesozoic reptiles. In the same vein Mantell had submitted a short paper to the Royal Society on the fossilised remains of the soft parts of molluscs. He rounded off this part of the discussion with an exhibition of microscopic specimens.

Lastly he discussed the tracks made by the fossil footprints found by James Deane and Edward Hitchcock in the Connecticut River Valley of Massachusetts. They were presented as made by the feet of giant prehistoric birds. Mantell had written to Professor Hitchcock:

> If this opinion be correct, it is clear that the feathered tribes of that ancient epoch were the iguanodons of their race, for the dimensions of one kind of footmark are fifteen inches long ... twice as large as those of the ostrich.

Despite assurances from Silliman and Lyell, Mantell remained sceptical. (They are now attributed to very early bipedal dinosaurs.)

In 1843 Mantell was in poor health with much suffering and a diminishing practice. His illness was organic and tangible but in addition to his worries he had the psychological burden of the attitude of Richard Owen, the aloofness of his old friend Sir Charles Lyell and the treachery of his protégé Richardson. Richardson's second edition was published by Wiley and Putnam of New York, shamefully pirating the works of Lyell, Whewell, Griffin, Lindley and Mantell without acknowledgment and impudently suggesting that his volume was an introduction to Mantell's *Wonders*. Friends advised Mantell to give up his practice altogether and to negotiate for a successor. As he told Silliman, he had a few thousand pounds at his command upon which he could subsist for two or three years, but no independent fortune. He planned two books which would add to the romance of geology and appeal to younger people. Not withstanding his illness, he attended the Geological Society in January and went with Reginald and Mr Bensted for the reading of his paper on Molluskites – found in the marine deposits at Maidstone – and returned home at 12.30 p.m. Illness did not prevent him from attending the council of the Geological Society, a lecture on cell-life and a soirée held by Lord Northampton, President of the Royal Society

On 11 March, he dined at the Clarendon Hotel with Mr Warburton, President of the Geological Society, before going to a splendid soirée (the last) 'of my friend the Marquess of Northampton, President of the Royal Society'. Prince Albert was there. Admiral Sir Edward Codrington introduced him to the Premier, Sir Robert Peel, who received him very courteously and accompanied him to the table on which was the model of the Temple of Jupiter Serapis. Sir Robert Peel listened with much attention and conversed with Mantell for nearly half an hour; but although courteous, Mantell felt that there was something repelling in his manner. It was said of Peel that his smile was like a glint of winter sunshine on the brass handles of a coffin. Mantell returned home dreadfully fatigued, suffering so much all the while that he had no pleasure in the gay scene.

Once again he wrote yet another epistle to the British Museum and to Sir Robert Inglis to complain that much of his collection had still not been unpacked. Sir Robert, a Tory MP, was notorious in Mantell's mind for having blocked the entrance of Queen Caroline to Westminster Abbey, preventing her from attending the coronation of her husband, George IV. The lack of news from Walter in New Zealand was a constant cause of worry, but he was greatly relieved when he received a copy of the New Zealand Gazette with an account of a lecture on geology given by Walter to the Mechanics Institution at Wellington. 'There is this satisfaction to be derived from the news, that Walter has not lost all taste for his earlier pursuits.' On hearing this news he sent Walter a good microscope and some books. To Silliman he wrote on the subject of brass rubbings made from the effigies and inscriptions on plates of brass from the time of the Norman Conquest, thereby recounting to him that the name Mantell occurs in the roll of Battle Abbey among the list of knights who accompanied William the Conqueror from Normandy. In the Mantell chancel in Heyford church, near where the Mantells settled, he had come across the following inscription:

> Jean Mantell gist icy
> Elizab, sa femme auci
> De lo' Almes Dieueit merci, Amen

He expanded further on the genealogy of the Mantell family, and also sent a copy of a paper read at the Geological Society 'Description of some Fossil Fruits from the chalk formation of the S.E. of England': *Zamia Sussexiensis* from a sandbank at Selmeston, *Abies Bensted* from the greensand near Maidstone, and *Carpolithes Smithiae* from the white chalk of Kent.

Increasingly, with escalating enthusiasm he took advantage of travel by rail to visit Winchester and to go to Matlock in Derbyshire with his son, visiting geological sites in the Peak district.

> The railroad in eight hours takes us within ten miles of that enchanting spot, and an hour's ride by coach completes the journey. Matlock is called the Gem of the Peak and Matlock vale is a fissure of some six miles in length and one and a half miles in breadth through mountain limestone adorned with quiet beauty. One side of the ravine is abrupt and the lovely Derwent murmurs along its base. The other rises to an altitude of 1,800 feet. Thermal springs issue out from fissures in the rocks and deposit tufa on everything over which they flow, and their ancient effects are shown in enormous beds of travertine The trap dykes and veins of Mt Linnest are of great interest and Crich Hill is an interesting example of elevation by protrusion of lava.

Later he went to Leicester with his son and daughter observing the strata of the Midlands with crags of granite, sienite, porphyry and slate. The children descended 700 feet down a coal mine and later visited Heyford in Northamptonshire, the Mantells ancestral home, where Ellen made drawings of the tombs and relics. He returned with a splendid fossil Cidaris from the Oolite of Wiltshire. In spite of his infirmities he enjoyed these geological and genealogical excursions with his children. Later that year he sailed by boat to visit Lady Mantell at Dover on the occasion of her eighty-fifth birthday. With these excursions his practice must have dwindled in spite of the assistant left in charge.

In writing to Professor Silliman he announced the death of their mutual friend Robert Bakewell and he described from the meeting of the British Association an upright trunk of a large Sigillaria found in a coalfield near Liverpool with massive roots and radicals containing Stigmaria ficoides

Throughout the year he had been working on his *Medals of Creation*, subtitled an Introduction to the study of Organic Remains. Joseph Dinkel had drawn the first of six major plates and Ellen had contributed 100 drawings to the thousand page text. The geological excursions which occupy the last hundred pages were originally intended for a separate volume but he was compelled to insert them to satisfy his bookseller. He intended dedicating it to Lyell in commemoration of their earlier friendship, notwithstanding his frigidity. Lyell failed even to acknowledge the dedication and Mantell wished he had made the dedication to Lord Northampton, Bakewell or Murchison who would have received the compliment cordially. It was eventually published at the end of the year in two volumes by Mr Bohn but, at such a cost, was unlikely to prove profitable.

The *Medals of Creation* was the first modern synthesis of paleontological knowledge written in English. He began with a theological, moralistic introduction and then gave a full survey of British Strata from the Drift of

Alluvium, ice ages, etc. to the Tertiary period with its divisions of Pliocene, Miocene and Eocene, to Secondary formations from chalk to sandstones and finally the Palaeozoic rocks of the Silurian and Cambrian periods. The discussion of fossils led into fossil zoology and the Age of Reptilia, and the last 100 pages were given over to geological excursions.

The year ended with Mantell's election as Fellow of the Royal College of Surgeons. He was among the first 300 to receive this honour. Reginald was about to seek an apprenticeship, and the Mantells went to London with Mr Pritchard, a geologist friend and author of *A Natural History of Animalcules* and *A History of Infusoria,* to see Isambard Kingdom Brunel for an interview to agree terms to place Reginald, who was nearly seventeen, with Brunel. Soon Reginald was employed upon a chain suspension bridge across the Thames near Westminster. This bridge, intended as the Hungerford Bridge, was afterwards removed and is now the Clifton suspension bridge over the River Avon. Reginald was involved in electroplating the enormous iron chains to prevent them from rusting. 'Reginald got on very well with the great engineer and promises to be very clever'. Mantell took him along to the soirée of the President of the Royal Society where they conversed with Herschel and Faraday.

On 3 June, Mantell dined with Andrew Pritchard and Sir John Herschel. Reginald arrived later with the Ross microscope to show Herschel infuroria. Pritchard, who sold microscopes, used a hydro-oxygen apparatus to display some fine specimens. Henceforth the microscope, and often Reginald, attended several soirées, e.g. Murchison's at the Geological Society as a prelude to yet another book, namely *Thoughts on Animalcules*, or *A Glimpse of the Invisible World Revealed by the Microscope*, dedicated to the Marquess of Northampton. After describing various species, Mantell opined that it is likely that many of the most serious maladies which afflict humanity are produced by peculiar states of invisible animalcule life.

In June Mantell visited Sir Robert Peel in order to meet the King of Saxony. To his surprise Peel, whom Mantell felt had a taste for Art but none for science, made some amends when he told Mantell at the reception that he had been reading *The Medals* that morning before breakfast with great delight. The King of Saxony went 'geologising' in Derbyshire after his arrival, making full use of Mantell's book. Mantell took an excursion with Reginald to visit the Isle of Wight, and discovered in the Wealden strata of the island's species of *Unio*, as large and as massive as are the splendid shells of Ohio and Mississippi, but was unwell on the trip and returned much fatigued. He also learnt that the New Governor of New Zealand had arrived, had received Walter very graciously and promised to promote him to some better and more profitable employment. Reginald, at seventeen, began his apprenticeship under Brunel with great advantages.

Chester Place, 1844-52, was Mantell's house in London after his move from Clapton.

having already been invited to the soirées of the President of the Royal Society and had met and conversed with Herschel, Faraday and others.

In August with his brother Thomas, Gideon Mantell went house-hunting around Eaton and Chester Square in Pimlico, ostensibly to provide a base for Reginald who was also involved in the development of the Great Western Railway. His lease on Clapham Common had expired, and having failed to sell his practice he hoped to continue in a consultative capacity. On 29 August he moved into No. 19 Chester Square with Reginald, his widowed sister Mary West (herself an invalid and eventually alcoholic), his assistant Mr Hamlin Lee (the son of Warren Lee of Lewes), one man servant, two female servants, a cat and a canary. The house was lavishly decorated with ornate carpets, and the busts, diagrams, fossil cases and artefacts of his previous houses. He reported to Silliman that his health was 'better than it was last year and the sufferings are not as great although any exertion and even a short walk is painful'. In a typical ranting mood he castigated Murchison: 'Would you believe it but Mr Murchison though living in the next square not 400 yards from my house has not yet even called on me.' Ellen Maria left home feeling that she had been treated poorly despite the work she had achieved with the drawings for *The Medals*. She became an illustrator for the publisher John Parker, whom she eventually married in February 1848. Mantell disapproved of the marriage and she remained distant from her father for eight years.

In October he delivered a lecture on the nervous system to the Clapham Athenaeum and later that month visited General Trevor and his daughter, Mrs Cotterall, at Glynde. On 14 November he delivered the first of six lectures on geology at the London Institution, which passed off very well with a good attendance despite the rain. At the third lecture there were 700-800 people present. Before the fifth lecture he had been ill all week, and the December weather was intensely cold affecting his neuralgia; even so it went off well. In the severe cold of January 1845 he visited a patient in Brighton and delivered the last lecture (on the nervous system) with the thermometer down twenty degrees below freezing to a crowded theatre. Also on 12 February he wrote to Archdeacon Robert Wilberforce, eldest son of the philanthropist. That night he had the most severe attack of neuralgia he had experienced since leaving Clapham and was confined to his bed for several days, fainting and injuring his face.

Despite his illness he wrote a short paper on 'Geology seen from Leith Hill' for his friend Brailey's *Topographical History of Surrey* and dined with Lyell, meeting Professor Kingsley and introducing him to Professor Buckland. In March he went to a soirée at the Royal Society, meeting Prince Albert and Sir Robert Peel, and made a short visit to the Isle of Wight observing a petrified forest with bones of Cetiosaurus and the stem of Clethraria. He wrote a pamphlet, *Memories of the Life of a Country Surgeon,* that May as a commentary on Sir John Graham's Medical Bill. (Quite what the bill was about is unclear; Graham had been First Lord of the Admiralty, had helped draft the reform bill of 1832 and was Home Secretary from 1841 under Peel but resigned in 1846).

He saw the SS *Great Britain* with Reginald but was too ill to attend the opening of Hungerford Bridge in which Reginald was involved. However, at the end of May he attended a soirée of Sir John Rennie, President of the Institute of Civil Engineers, meeting eminent engineers such as Brunel, Walker, Cubitt, and Leopald von Buch. In June he visited Lewes, going in a fly with his brother and Mr Warren Lee to Southerham, Bridgwick and Offham chalk pits. This jaunt later led to an article entitled 'Rambles through Lewes'. He also bought a slab of Sussex marble and arranged for the selling of his house at Castle Place, which had been rented out in November.

*

Mantell's microscopic researches came to fruition in May and June 1845 in the form of two papers read before the Geological Society and the Royal Society, respectively, and later published. The first was entitled: 'Notes of a Microscoptical Examination of the Chalk and Flint of the South East of England; with remarks on the Animalculites of certain Tertiary and Modern Deposits.' Even when paraphrased the introduction is dramatic:

> The founders of the Society could scarcely have imagined that the structures of minute forms of animal existence invisible to the unassisted eye would become a legitimate subject of geological investigation and that the coverings of their miniatures of life would be preserved in a fossil state constituting a large proportion of many rocks of great thickness and extent; still less that the soft perishable bodies of animalcules could be preserved by mineralization, and be found entombed , like flies in amber, in the flint nodules of which our roads are so largely constructed.

The Prussian microscopist, Christian Gottfried Ehrenberg paid a lengthy visit to Britain in 1838 with a series of collecting trips to southeast England and Norfolk. He lectured to the British Association and the Geological Society. He discovered that thin slices of flints showed evidence of dinoflagellates and other bodies which he considered to be thick walled fungal spores and freshwater algae (*Xanthidia*). Bowerbank and Wilkinson reported the presence of sponges within flints. There had also been earlier reports by Lonsdale and Parkinson that various kinds of sponges and alcyonia had become enveloped and saturated in silica. Mantell was also inspired by his friends, the Rev. Joseph Reade and Henry Hopley White (who had presented him with the microscope designed by Andrew Ross), to examine the mircofossils in white chalk and limestone. With his son Reginald and Hamlin Lee, he found *Rotaliae* and other animalcules present as silicaeous shells, the type of organisms varying with the strata: chalk, limestone or oolite.

In his second presentation, he observed that the shells of many Rotaliae were filled with a substance, varying in appearance from a dark opaque brown, to a light transparent amber, resembling in form, the soft bodies of existing species of Polythalamia. Helped by a chemist, Henry Deane, he used weak hydrochloric acid to remove the outer calcareous shells leaving a residue of quartz, silicate of iron and numerous remains of the soft parts of animalcules, comparable to those of living Roataliae. The presentation ended with a series of drawings displaying their complex structures.

*

In the summer of 1845 the new railway line from Lewes to Brighton was driven through the ancient site of Lewes Priory. One outcome of the despoliation of the Priory, which was dedicated to Saint Pancras, was the unearthing of the leaden cists of William de Warrenne and his wife Gundrada. (It has been said that William's wealth, scattered widely over many regions of the country, would have outshone that of Bill Gates.) The railway directors, faced with these pious antiquities of the Norman period, offered to lend 'every possible assistance' and provided £50 for a suitable receptacle for the relics as long as they were assured that they would always be open to public view in Southover church. 'The unexpected bounty of the iron horse convinced William Figg that the common idea that railways are destructive in their tendencies and can be of little use to archaeology or the arts was simply wrong'. The find brought leisured tourists to Lewes; thus M. A. Lower, the Sussex historian, declared: 'Our town is rapidly rising with greater celebrity than it has ever yet enjoyed.'

He was intrigued by the balloon flights from Battersea of Mr Green, and although he had travelled by railway on many occasions wrote that:

> Millions spent on the railways will not be required, sooner or later travel from aero-stations will entirely supersede railway travel. We have had a frightening number of accidents lately, no wonder, the express trains which go at a rate of 50 mph are terrific to behold. They raise up clouds of dust induced by their passage through the air.

Silliman was far from impressed: 'I do not see with you how we are ever to travel in balloons except before the wind; they have never been adequately steered and it appears to me demonstrable that they never can be.'

For once he records praise in his journal:

> Mr Lee left today having obtained the Curatorship of the College of Surgeons of Edinburgh. He has been with me ever since I left Brighton. He has been all through a deserving man.

In September Mantell went to the Isle of Wight, finding a femur of an Iguanodon 3 feet 4 inches long, vertebrae and one tooth as well as fossil fishes. He read a paper on the Wealden strata of the Isle of Wight at the Geological Society; unfortunately while in transit in the carriage to the Society the thigh bone broke. The excursion to the island led to a book: *Notes on the Wealden strata of the Isle of Wight with an account of the bones of the Iguanodon and other reptiles discovered at Brook Point and Sandown Bay*.

In his first letter of 1846, Silliman, from afar and never having met, had to write to cheer Mantell up and disabuse his mind of the existence of imaginary enemies. In fact Mantell felt desperately lonely and, to his disappointment, found at the annual meeting of the Geological Society that he was not elected as Vice-President nor re-elected to the council. He attributed this, probably correctly, to his being too independent, and because he could not sanction many of the proceedings of the ruling coterie, though he respected them individually. But he was pleased to attend the President's Reception at the Royal Society with his son Reginald at the invitation of Lord Northampton. He took along a model of the Isle of Wight and spent some time discussing it with Prince Albert. Sir Robert Peel was there looking very ill and careworn; 'and well he may. He had had to withstand the virulence of the Landed gentry against his measures to repeal the Corn Laws' – presumably comments such as Mantell's that the income tax ought to damn the everlasting fame of Sir Robert Peel. Also present was Sir Roderick Murchison, back from Russia and just knighted. From envy perhaps, Mantell noted that he may wear the red sash, cross and star of the Order of the Emperor of Russia bestowed upon him.

In February Mantell lectured to 500 people at the Clapham Common Athenaeum on 'the Elementary Principles of Physiology'. He also returned to Clapham to lecture on brass rubbing and mesmerism. For this he received a vote of thanks:

> The sincere thanks of this Society be offered to Dr Mantell for the lecture he delivered this evening and feeling that the original function of the

> Society had been mainly caused by the exertions of Dr Mantell for the furtherance of scientific pursuits in the neighbourhood they beg to express to him the satisfaction they feel in finding that the removal of his residence from the immediate neighbourhood will not prevent him affording his valuable assistance in future.

Writing four days later, he gave Professor Silliman an account of the meeting and the reasons for the change he had made.

> His son was now working for Brunel needed to be within a short distance of his office; the termination of the lease on his house; and the hope of starting a consultative practice which would be personally less fatiguing.

From his pretty and cheerful house, on a twenty-one year lease in the heart of the most aristocratic part of London, close to Belgrave Square and the Queen's Palace, he was engaged to deliver six lectures on geology, being paid the high fee of £10 10*s* a lecture of one hour. Institutional lectures formed a relatively secure source of income on a contract basis (thus Humphry Davy received a salary of £500), and were in part dictated by the need to present popular science at minimum cost. By this means he was able to supplement his meagre earning from other sources. He had not been able to sell his practice, but would return to such of his patients at Clapham still left, hoping by the time he lost all the remainder he would have other means to cover his expenses. Lyell called directly he returned to town and heard the new address, and he also attended the first of Mantell's course of lectures at the London Institution. The room held nearly 1,000 people and was crowded.

The move made him more aware of political events, and he began to feel the desolation of living in a crowded city. He contrasted the superb equipage and splendid liveries of the rich with the discontent of the lower middle classes, charterism, the slaughter of our armies in India and the potato blight, affecting Ireland in particular. He quoted Cobbett that the Solanum would prove a national curse, from the habits of idleness it permitted, and should be replaced by more wholesome and less precarious vegetables. He dedicated his most recent publication, *Thoughts on Animalcules, or A Glimpse of the Invisible World Revealed by the Microscope*, to Lord Northampton and sent copies to Peel, Northampton, Faraday (whose lecture on magnetism he attended), Parish, James Clark, Mr Lee, Miss Foster and Mrs Williamson. In its conclusion he refers to the distinguished clinician, Henry Holland, who in his *Medical Notes and Reflections* (1839) had a chapter on The Hypothesis of Insect Life as a Cause of Disease. In agreement with Holland, Mantell wrote it is probable that many of the most serious maladies which afflict humanity are produced by states of invisible animalcule life.

In June he paid a visit to Banwell, the grounds of the Bishop of Bath and Wells, on the steep slope of a hill, noted for the deposits of various bones of deer, ox and wolf found there. Sir Robert Peel resigned at the end of that month.

> So much for the country, but the people are heartily sick of Whig, Tory, radical, protectionist and all other political charlatans! If they do not mind we shall have a grand row that is certain murmurs, not loud, but deep are heard – everyone against the proceedings of the House of Lords and the scandalous neglect of public business which party and personal squabbles have occasioned.

On 29 June he journeyed by rail to Brighton, and thence on the newly opened link to Lewes railway station and was met by his brother and Warren Lee. He visited Castle Place, which was still unoccupied, and then travelled to Hamsey to collect fossils. Mr Grantham called from Barcombe. The next day he went by phaeton to Cooksbridge to cuttings on the chalk marl near Keymer to obtain fossils from the railway workers ('or navies as they are now called'). He used his visit to write *A Day's Ramble in and about the Ancient Town of Lewes* (150 pages). The frontispiece was a drawing he had done when aged eleven of the dovecot at Lewes Priory. He described Southover and St Michael's churches, the new church of St John's Subcastro, Anne of Cleves' house, the High Street, the Bull Inn where Paine lived, the Castle Keep, Alban Street where was the Lewes library society and John Button's Academy at Cliffe. He hoped the book would help him sell Castle Place.

On 18 October Mr Reade paid him a visit and they sat up till 1.00 a.m. over the microscope with Mantell convincing his guest of the presence of the soft parts of rotatiae in the chalk. The day before, between five and six o'clock, Dr Fitton – who had often visited Mantell partaking of his hospitality at Lewes, and for whom he had often laboured all night to compare his fossils, or answered his interrogation respecting the geology of southeast England – called upon Mantell for the first time since he came to reside near London. Fitton told Mantell he should not have called had he not heard that Mantell was very ill, and after some common remarks broke out in a vehement storm of reproach on his 'illiberal conduct in Science' and said he considered the spirit of monopoly, which Mantell carried into everything he engaged in, was highly reprehensible and most inimical to the progress of science.

A painting by Gideon Mantell of the west view and gateway of the ruins of Lewes Priory, 1809.

I never experienced more astonishment than at this unexpected and most unjust accusation. I defied him to allege a single proof either from my conduct or in my writings. I conjured him to point out one sentence or even a line that would bear such interpretation; but this he would not do, but repeated the allegation and said that many other persons thought as he did; and that was the reason why he had called upon me. He took his leave saying he should call again and hoped I would not be hurt at what he had said; but he was always open and candid to his friends!!! Were this man not an Irish man, I should think him demented.

His object in calling upon me now I cannot divine! And this man (as my works prove) I have taken more pains to treat with all respect for his geological researches, than any others; any never willingly behaved towards him but in the most courteous manner. I can now understand some of the unjust conduct I have experienced from the Geological Society; the non-publication, and actually the loss of my paper on molluskite and its illustrations; the refusal to consider my paper on Fossil Reptiles published in the Philos. Trans. as a candidate for the Royal

> Medal for that year, when it was the only paper; and Dr Fitton who was on the committee objected to it because it was not on Geology but on Palaeontology. But it is folly to dwell on these petty annoyances:
>
> Do I regret that roseate youth has flown
> In the hard labour grudg'd its niggard meed
> And cull'd from far and juster lands alone
> Few flowers from many a seed. No!

In October the Geological Society refused to consider his paper on fossil reptiles published in *Philosophical Transactions* as a candidate for the Royal Medal for that year. This did not stop him attending an evening party with Lyell and Murchison where they discussed the planet, newly discovered by John Couch Adams in 1841, Le Verrier (Neptune). But publication was anticipated by Urbain Le Verrier. William Herschel (father of Sir John Herschel) had discovered Uranus in 1781; the English and French astronomers had predicted its site, and Johann Galle, from Germany, made the discovery in September 1846. Mantell could not resist writing to Silliman about the event.

> Adams is a modest, unassuming young man, the true discoverer of the new planet. He had actually calculated the planet's magnitude, orbit and place months before Le Verrier even guessed that the perturbation of Uranus might depend on a much more distant planetary body and actually deposited his calculations with Airy.

Sir George Airy was the Astronomer Royal for some forty years. (In 1980 a Galileo scholar announced that Galileo himself had first seen Neptune in January 1613. Galileo's sketches showed its apparent motion, but his interest was concentrated upon Jupiter and its satellites.)

He attended the anniversary of the Royal Society on 30 November:

> All passed off quietly. Copley Medal to Le Verrier for his planet!
> Royal Medals to Faraday for his Magnetism paper!
> And another Royal Medal to Prof. Owen for his paper on Belemnites!!!
> – a paper which turns out unfortunately to be a tissue of blunders from beginning to end!
> So much for Medalists!
> Came home to dinner, and worked on the Isle of Wight. A severe frost suddenly set in; the thermometer last night in the Regent's Park down to twenty-five degrees; colder than at any time last winter.

The cold and dampness continued through December, bringing on an intense bout of sciatica. He took acid hydrocy, which failed to relieve the pain. Neither did three minims of 'Tincture Opii' – which made him sick. The attack subsided overnight but he felt ill all the next day.

William Smith had noted the fossils in the various strata when developing the canal system. Also the building of the railways, particularly it seems The Great Western Railway, unearthed belemnites and other marine fossils. Professor Owen was keen to label a new species *Belemnites owenii* as part of his desire to raise himself in the public esteem as the supreme, the Newton of Natural History and to lobby the Prime Minister, Sir Robert Peel, to found a National Natural History Museum. He had chaired the meeting at which his own paper on Belemnites, previously communicated to the Royal Society, was recommended for the prestigious Royal Medal. Though Owen received the medal, Edward Charlesworth, the editor of the *London Geological Journal*, condemned his failure to acknowledge the earlier work of Mr Joseph Chaning Pearce, a railway engineer and amateur geologist who had read his paper to the Geological Society describing the new fossil not as a Belemnite but as belonging to another genus which he had named *Belemnoteuthis*. Its body was composed of fifty chambers and had an ink sac and ten arms with pairs of hooks and suckers. Reginald Mantell, when he returned from America, was also working on the Great Western Railway under Brunel, and between Chippenham and Trowbridge his team uncovered superb fossil Belemnites. In 1847 he sent his father specimens confirming Pearce's observations. Owen had falsely ascribed features to the Belemnite which it did not possess. Mantell, after investigating the question, sent a paper to the Royal Society but felt much reluctance in proving that Owen was in error. This was to lead to yet another stormy confrontation with an irate and vindictive Professor Owen. Even at the College of Surgeons, senior members were complaining that Owen was 'proud, imperious, overbearing, selfish and jealous, especially towards his rivals and juniors.'

January 1847 brought a grumble about the price of coal, heightened as the result of severe snow in the north. In London thousands skated on the Serpentine and Mantell was laid up with a severe bout of influenza, suffering as usual both in mind and body. The cold spell lasted for eleven weeks. However, on 13 February, he was able to attend a soirée of the Marquess of Northampton. Among the 500 people present he met Mr Adams 'the true discoverer of the new planet' and Sir Robert Peel, 'as cold and unapproachable as ever.' Yet ten days later he was pleased to accept an invitation from Sir Robert Peel to see his pictures. The afternoon proved very favourable for the inspection. There was a select company of nobility, savants, and artists. Sir Robert received him very graciously.

> The pictures were very fine; some drawings by Rubens were marvellous productions. It was a very great treat indeed. He met many persons he knew.

Publication of the 'Isle of Wight' was going ahead with a long title: *Geological Excursions round the Isle of Wight and along the adjacent coast of Dorsetshire, illustrative of the most interesting geological phenomena and organic remains*. Below the title was a fine engraving of a Fossil Lobster from Atherfield. The inscription was to His Royal Highness Prince Albert as a testimony of admiration and respect for his liberal encouragement of British Science and Art – this by permission – but in Mantell's opinion with insufficient appreciation. He began the journey from London detailing the various strata through which a train would pass. The first ten-twenty miles from London traversed beds of clay, loam and loosely aggregated sand and gravel. The next geographical unit was the chalk downland, invariably traversed by steep cuttings and tunnels. Beyond the downland, a whole new series of older clays and sand to be followed by Jurassic limestones on the Bristol and Birmingham lines and by the Wealden beds to Brighton. In the next section he gave lists of various excursions and voyages, acknowledging the works of former authors and including visits to Swanage, the Isle of Purbeck, the Isle of Portland and along the Dorset coast. The book by English standards was financially successful, but nothing like the amounts paid to American authors.

The major work of the year was the extensively rewritten, considerably sophisticated sixth edition of *The Wonders of Geology*, dedicated to Professor Silliman. Mantell refers to it as his lucubrations – 'laborious study, especially done at night!' There were still two volumes and eight talks, but several were recast in two parts. His sketch of the nebular theory now included an important footnote tentatively endorsing the formation of living beings from inorganic elements. He followed Lyell in preferring gradualism rather than a series of catastrophes, and excluded the Deluge as a geological agent – but later wavered in this regard. He noted that whole races of animals sometimes disappeared from strata undisturbed by physical catastrophes, as in Russia. He was now aware of geysers, geology and wingless birds in New Zealand as well as elsewhere. His chronological synopsis of rocks and strata was further updated with reference to Lyell and Murchison. Along with Dr Grant (1835) he stressed the anatomical proximity of monkeys to man and expanded his researches into microfossils.

He amended his ordering of fossil reptiles to approximate more closely to that of Owen, but felt that the Greek root *deinos* as in *Dinotherium* and *Dinornis* (terrible beast and bird) would mislead the unscientific by suggesting affinities between these two unrelated creatures. Buckland received full credit for Megalosaurus, a carnivore thirty feet long. Fresh details were added to his description of the Iguanodon, and the Hylaeosaurus, twenty-thirty feet long with its dermal spines, was basically unchanged. Darwin's 'delightful volume' on the structure and distribution of coral reefs added to his discussion of corals and crinoidea.

He found the task of bringing the *Wonders* up to date, adding new data and acknowledging all those who had contributed to it, hard and weary, suffering, anxious and profitless toil.

> Although heartily sick of my task, I shall not regret it when the task is over, as it will have given me the opportunity of stating my opinions and defending my claims where unjustly set aside and overlooked...The meeting of the British Association at Oxford is just over; it has been a splendid one. All the elite have been loaded with academic honours. I did not go near for I felt unable to bear the excitement and besides I have neither friends nor interest. I must be content to throw a few pebbles into the ocean of truth, and pass away from this scene of trial and suffering unremembered and unregretted save by a few valued friends.

On 1 May he went to St Bartholomew's Hospital to witness two operations under ether. He commented that the loss of sensation in both instances was complete, with no consciousness of the operation but the effect on the eyes was appalling. He saw Jenny Lind perform Norma at the opera in June. And in September visited Lewes, having managed to sell his house for £950, and giving a lecture on geology in the Lewes Mechanics Institution. One outcome of his involvement with the Isle of Wight came in November when some fishermen sent him several vertebrae and some teeth from an Iguanodon. One of the teeth was so perfect and unworn as to indicate continual renovation.

He had good news from both sons. Walter was sending a large collection of Moa bones from New Zealand and Mr Brunel had cancelled Reginald's articles, launching his career at the age of twenty and giving him a salary of £175 per annum for the last year, besides the keep of his horse. He left a letter to thank Mr Brunel for his liberality to his son. In December, Walter's box arrived with more than 800 specimens in fine preservation, containing many novelties.

On 25 February he dined at Sir Robert Peel's as part of a party of sixteen gentlemen, Lady Peel being the only lady. The Dean of Westminster (Rev. Buckland), Chevalier Bunsen, Sir James Graham, Mt Hallam, Sir H. de la Beche, Mr Lyell, and Prof. Owen were of the party. He took his microscope and showed a few specimens to Lady and Miss Peel.

> A very pleasant evening: Sir Robert most gracious and unbending. The French Revolution was the engrossing topic; every half hour bringing some fresh rumours. When he left at 11.30 Sir Robert expressed his gratification with his new edition of the Wonders, and said he was delighted with the "Isle of Wight" which he had twice read through.

The political excitement of the French Revolution was once again the main topic at a soirée of the Marquess of Northampton held a few days later and attended by Reginald, Sir Robert Peel, The Dean of Westminster, the Bishop of Oxford (the Rev. Samuel Wilberforce) and that of Norwich.

Professor Owen attacked Mantell most virulently at the Royal Society on 23 March when Mantell's paper on Belemnites was read. Mantell had himself discovered that the guard of the Belemnite was invested with a shelly capsule which, extending upwards, formed the receptacle. This fact was previously unknown. Reginald's specimens showed that the upper part of the phragmocone had elongated shelly processes, probably for the attachment of muscles which moved the cephalic arms. After the paper had been read, Owen arose and for half an hour made the most ungentlemanly and uncalled for attack, ridiculing Mantell for wasting the time of the Royal Society and denying the higher degree of accuracy in observation and deduction shown by Mantell. In the face of this onslaught both the Dean of Westminster and the Marquess of Northampton rallied in support of Mantell. The second issue of the *London Geological Journal*, in which Mantell contributed a brief account of *Unio valdensis*. Charlesworth repeatedly attacked Owen, emphasising the dangers of 'official' opinion and the desirability of maintaining open scientific debate.

On 25 March 1848, Captain Lambart Brickenden of Warminglid, Sussex, the current owner of the quarries at Whiteman's Green, sent Mantell a twenty inch specimen, part of an Iguanodon's lower jaw. Mantell accepted the gift in exchange for a set of his own works (viz: *Medals*, *Wonders*, *Lewes*, *Animalcules* and *Pebble*). He took the finds from the Isle of Wight and Tilgate forest to Dr Melville[2] at the British Museum for comparison with other fossil and recent reptiles, later exhibiting the specimens at the Royal Society on 25 May. In the finale to his presentation, he re-identified a specimen formerly presented in 1841 as part of the lower jaw of a young Iguanodon as sufficiently distinct to deserve recognition as a different species which he called *Regnosaurus Northamptoni*. No other specimen of Regnosaurus has been identified,

To celebrate Reginald's twenty-first birthday at the beginning of June, Mantell went to Bath, thence by canal boat to Bradford and walked to Trowbridge with Reginald. They visited Burfield quarry and the ruined mansion of the Duchess of Kingston before returning to the inn on a thoroughly wet and stormy night. Back in Bath they saw the private collection of the late Mr Chaning Pearce with its perfection of Apiocrinites and other crinoidea. The next day at the Geological Society he read his paper on the Wealden and exhibited his jaw bone of the Iguanodon to a very full meeting, at which he met up with Captain Lambart Brickenden in person.

In June he went to Lewes to see Mrs Thomas Mantell who was very ill and died the next month, a week after an amputation of her leg. About that time he heard of Richardson's suicide. On 1 July he dined at the Athenaeum and went to the Haymarket theatre and saw Mr and Mrs Kean in *The Wife's Secret*. He had never seen Mrs Kean before.

> She is incomparably the best actress on the English stage. Kean is much altered, his manner being more prominent and his voice very rough, and at times absurdly pedantic; but he is still as earnest and the best actor we have; better unquestionably than Macready.

On Sunday 23 July, he breakfasted with Mr B. D. Silliman, son of his friend, whom he found delightful, amusing and intelligent.

In August he spent an evening at Covent Garden hoping to see the opera *Les Huguenots* but was refused admittance with great insolence because he wore light-coloured trousers instead of black, yet the same sartorial apparel as he had worn to the Marquess of Northampton's dress soirées. He expected that the door-keeper wanted a bribe but he refused. The next week he left for Bath and went with Reginald to the British Association meeting at Swansea. He visited various country houses, even though the experience was punctuated by heavy rain In September he did a lecture tour of the Isle of Wight, and in October lectured in Brighton. Lord Gage of Firle attended both days as did Mrs Allnutt.

His old friend Lyell had just

> allowed himself to be knighted. It is the old story of the Fox and the Grapes. I shall soon be the only old philosopher who can boast of being 'Unpaid, Unpensioned, and Untitled'[3]

Benjamin Silliman Snr was probably amused as he made a delicate reply.

> As a good republican he thought that a man, eminent by his own qualifications and doings, was not exalted by a title "although in a country of monarchical and autocratical institutions it may give him a degree of adventitious consideration."

Working in conjunction with Dr Melville, Mantell was able to establish the relationship of the upper and lower jaws and teeth of the Iguanodon and received the Royal Society's gold medal as a result. Together they went to Horsham to look at Mr Holmes' Wealden fossils. Mantell was much gratified by his trip, for though Mr Holmes would not allow him to make drawings or notes of any specimens, they obtained a good idea of his collection and ascertained the imperfect data on which many of Prof. Owen's determinations of the Wealden reptiles were founded. A further find were bones of the Iguanodon from the Weald of the south side of the Isle of Wight, more colossal than any previously known. A femur was twenty-eight inches in circumference and a tibia of eighteen inches. He also obtained some glorious specimens of fossil foraminifera – the soft parts – from the chalk. The same phenomena had been observed in the limestone of India. 'It appears that the soft bodies have been detected as numerously as in our chalk yet our *philosophosophists* laughed at my first announcement of this curious fact; now it is denied by no one.'

In January 1849, George Scharf was able to complete drawings of the vertebrae and other bones of the Iguanodon, enabling Mantell to reconstruct the overall appearance of the animal.

1849 began with lectures by Faraday on magnetic-crystalline action and on magnetism; the latter at the Royal Institution before Prince Albert. Sir Humphry Davy received £500 per year for his institutional lectures, but the exemplary scientific lecturer was not Davy but Michael Faraday with his clarity, neatness, arrangement and concentration on the subject to hand. Mantell had lectured on 'A Frog and a Pebble' at Clapham Athenaeum and attended meetings of the Antiquarian and Royal Society in which William Parsons, 3rd Earl of Rosse[4], was in the chair and spoke on spiral nebulae. Later they discussed his mighty telescope. Mantell's paper on which he had worked with Melville was delayed till the March meeting when the library table was covered with specimens of Iguanodon and Hylaeosaurus – sacrum, humerus, femora, vertebrae, scapulae, coracoids, tarsals and metatarsals. Professor Owen did not attend, but Lord Northampton and Sir R. Murchison commented with observations on the labour involved.

In March he was horrified when Reginald went to Bristol on the engine of the express train. A man was crushed to atoms by the engine at Swindon. Reginald had been given the task on riding aboard several of Brunel's new and decidedly unsafe locomotives on a frightful series of test runs to ascertain their consumption of coke and water. Mantell was unprepared for him to continue that dangerous occupation and paid for him to tour America. Just before Reginald was preparing to visit America, he went with his father to Earl Rosse's soirée at Somerset House. The surgeon, Mr Alfred Smee had some very fine injections of the brain (mainly of a cat) under the microscope which Mantell inspected along with Sir Robert Peel and M. Guissot. On 9 May, Thomas Mantell accompanied his nephew to Euston Station to begin his trip on the steamer *Caledonia*. Mantell consoled himself after his son's departure with a series of lectures in Brighton attended by over 400 people, including Sir Woodbine Parish, Mr Ricardo and Mr Grantham with his son.

The absence of his family worsened his gloom with headaches and spasms in his chest for which he took chloroform, to no avail, though later that day he went after dressing with difficulty to dinner and a delightful evening at Sir Robert Peel's. Later that month he assisted Mr Lawrence and Mr Stanley in an operation removing a large portion of the left ramus of the lower jaw of a Mrs Steer who had taken lodgings near by. Chloroform was administered with some difficulty, the operation was long and severe but with little consciousness on her part and the wound closed with sutures. She passed a good night and the next day could swallow without difficulty. He followed her progress with Mr Lawrence; on 14 June they cauterised the malignant fungoides (probably sarcoma of bone) but she continued to deteriorate and died on 16 September. Mantell also recorded other deaths: Sir W. Pearson from whom he had bought the practice, his friend Horace Smith, poet and author of Brambletye House and Mr Grantham's daughter. His practice became busy due to an outbreak of cholera, the spasms of which he treated with chloroform and calomel. By the end of September he was able to claim that all his patients with cholera were convalescent. Dr Buckland, as Dean of Westminster, gave a sermon at a thanksgiving service for the decline of the cholera epidemic with the text, 'Wash and be Clean'. His own illness with severe paroxysms of neuralgia of the right thigh he tried to alleviate with cold and hot applications, prussic acid liniments, ether, opium and calomel and inhaled chloroform twice.

Mr Fowlstone of Ryde brought him the head of a tibia of Iguanodon from Sandown Bay of enormous size, fifty-eight inches in circumference and portions of two unknown bones. His expenses and the sums paid to the men who found them and their carriage came to two pounds, which is much too high a price. When he went to Oxford he proceeded to Dr Buckland's Museum at the Clarendon and spent four hours examining the enormous collection of bones. 'His collection is enormous, crowded to excess but excites very little interest and his class has just six or seven persons'.

He was equally successful on a visit to Lewes. After he had called on his old friend Mr Grover, he walked with him to Mr Fuller, the miller at Malling Hill, to see a large bone from Cuckfield, a humerus fifty-four inches long, the extremities not perfect. It did not possess all the characteristics of the humerus of the Iguanodon, and he inferred that it belonged to the same species of gigantic Saurian as the four large lumbar vertebrae from the same pit, called Cetiosaurus by Professor Owen. Mr Fuller had the dermal spine of Hylaeosaurus, which he let Mantell draw. After a night of intense agony when the spasms of the nerves of the thighs came on as bad as ever, he inhaled chloroform and was insensible for half an hour. With no relief he got up at eight o'clock; at eleven he called on Grover and managed to crawl up the steep slope of Cliff Hill. Once he reached the brow of the cliff over South Street chalk pit, he laid on the turf and spent an hour and a half on melancholy musings over the past. He took in the view tracing the Newhaven train from Piddinghoe to its terminus at Lewes before descending and sauntering over to Southover church to visit the Warrenne tombs (To get up Cliff Hill with its steep incline after the agony of the previous night shows considerable determination.)

He went to an evening meeting of the microscopic society with Mr Alfred Woodhouse where the President, Mr Busk, read a paper on the supposed finding of fungi from the body of a person who died of cholera (Vibrio cholera). He then rushed to Lewes where Mr Grantham, his long time friend and co-sponsor of the visit of William IV to Lewes was dying, an event which caused Mantell to make his will. A younger geologist, Frederick Dixon of Worthing, who had been assiduously collecting the fossils from Lewes and Bracklesham ever since Mantell left Brighton, died presumably from cholera after being ordered to superintend the emptying and repairing a sewer belonging to some houses of his Westminster property.

Mantell himself delivered a lecture to nearly 500 people in Brighton on the *Invisible World of Being*, revealed by the microscope. Mr Ricardo, Sir Roderick Murchison, Basevi and Masquerier were among his old friends who attended. He took the artist Mr Dinkel to Mr Fuller at Malling Mill to make a correct drawing of the large humerus and offered seven guineas for it, but this was declined. Later he received a letter from Mr Grover to say he had purchased it for the same amount plus copies of *Wonders* and *Medals*. Then he settled to the task of writing a memoir on his new humerus from Tilgate forest which he intended to call Pelorosaurus (monstrously gigantic). Adding to the excitement of the occasion, he also received two boxes of Moa bones and many beautiful tertiary shells from Walter.

Mantell's activities belied the continuous restatement in his letters of his physical infirmities. In the *Geological Journal* he gave a full account of the anniversary dinner of the Geological Society and sent copies to his friends.

> The Archbishop of Canterbury, Sir Robert Peel, and the Russian Ambassador, were there; and my friend Sir Charles Lyell, the new President, took the chair. Murchison, De la Beche, Buckland, Sedgwick, and almost all our great men, were present. The Archbishop made an admirable speech in defence of Scientific pursuits and Geological in particular; and Sir Robert a senatorial declamation in like spirit. Lyell spoke good sense, but was so long in his pauses, and so hesitating, that I was frightened out of my wits lest he should break down. Dr Buckland made an academical oration, like one got by heart by a young collegiate; and Sedgwick poured forth a flood of eloquence, which, in spite of the disordered tones in which it was uttered (for his voice is most harsh), carried everything before it. The Belgian Ambassador, in capital English, with just sufficient foreign accent to add to its interest, gave a luminous address in praise of science, and in just encomoniums on his own country for having remained unmoved in the midst of the revolutionary tempest which had swept over the Continent. Murchison made a courtly speech, highly complementary to the nobles present; and your humble servant, who had to respond as one of the Vice-Presidents, gave a flourish of trumpets, which concluded the entertainment. Sir H. De la Beche has been a capital President; his address I hear was excellent.

Nowadays one could accept such a report as a private communication; as a more public one it would cause considerable irritation. (One would like to know how it was received in 1849).

FIFTIES

The year of 1849 ended with a monumental set-to showing Professor Owen at his most malign, presumably seeking revenge for the rubbishing of his paper on Belemnites for which he had been given the Royal Medal the previous year. He was intent on denying Mantell the medal in the following year. On 16 November Mantell attended a meeting of the Geological Committee of the Royal Society at Somerset House. Messrs Horner, Greenough, Darwin and Prof. Sedgwick were present. As his claim had to be considered, Mantell retired and left another member to officiate as secretary. Two days later he called on M. Gassiott and learnt that the committee had recommended Prof. Forbes papers on glaciers for the Royal Medal, and that his memoirs (chiefly from Prof. Owen's remarks) were considered unworthy of any recommendation. Very much harassed, he wrote to Professor Buckland, now the Dean of Westminster, remonstrating against the decision of the Geological Committee. The Dean sent a kind answer, expressing concurrence with Mantell's opinion and promising to do everything in his power to amend the decision. Mantell then sent a letter addressed to the President and Council complaining of the injustice of the Resolution of the Geological Committee and requesting further consideration of the question.

Canvassing opinion, Mantell firstly called on M. Gassiott, whose kindness and perseverance in endeavouring to assert Mantell's claims to the Royal Medal were beyond praise, to find that Prof. Owen had done everything in his power to prevent his obtaining it. The Dean of Westminster called and stayed two hours, and wrote a very strong recommendation to the Committee for his claims stating that all of Mantell's papers on the Iguanodon, on foraminifera (marine organisms with perforated shells; Mantell had treated flint, silicate of iron and chalk with acids to reveal the soft bodies of rotaliae and other foraminifera in an extraordinary state of preservation), and on belemnites, qualified him for the highest honours the Society could bestow.

Sir R. J. Murchison was as ardent and generous as usual for Mantell. That evening after a long debate, owing to the obstinacy of Mr Greenough and the infernal insinuations of Prof. Owen, who though not on the committee, had attended as a visitor to endeavour to stifle Mantell's claims, had decided in recommending the award to Mantell; but they would have to wait till the full council meeting. Murchison, Lyell, Horner and Prof. Bell were warm in Mantell's favour and Dr Buckland's letter was read.

Sir Charles Lyell wrote to Mantell. He wanted to be fully prepared with notes etc. should Prof. Owen's opposition be renewed. At the meeting Lyell discussed the merits of Mantell's studies, reminded Owen how often he had used Mantell's research in his own work, and quoted the high praise Mantell had received from Cuvier. On 30 November he was able to inform Mantell at Somerset House that after a long discussion the Medal had been awarded to Mantell with Prof. Owen and one other (probably Sir Robert Inglis) voting against. The Copley Medal was presented to Sir Roderick Murchison for his works on the 'Silurian Region, Geology of Russia, etc'; to Col. Sabine for his magnetic observations; and to Mantell for the memoir on the maxillary and dental organs of the Iguanodon. Mantell took meticulous trouble to thank both M. Gassiott and Lyell.

The obstetrician, Dr Robert Lee, called on Mantell to congratulate him on his triumph over Owen. On a similar occasion, Richard Owen had succeeded in preventing Dr Lee from receiving a Royal Medal for his paper on the 'Nerves of the Uterus', though worthy of receipt. When Mantell was informed of the decision and invited into the meeting room, Owen sat opposite him, looking the picture of malevolence. At a later meeting of the Royal Society, Owen came up and shook hands with people near Mantell, then stretched out his hand to Mantell, saying what a pleasure it was to see him there. Mantell bowed and declined the handshake.

To add a further episode to the saga of Owen's vendetta, the following year he applied to the Royal Society for permission to take impressions from illustrations of fossil reptiles published in the Society's journal.

> Professor Owen begs respectfully to request permission of the President and Council of the Royal Society to have 350 impressions taken from eight plates of Fossil Reptilia, described by him in his Report on British Fossil Reptiles, for the purpose of the illustrated edition of his work on that subject. He pledges himself to superintend the careful printing of the impression, and that the stones be restored to their original conditions and the aid thus afforded by the Royal Society will be gratefully acknowledged for the work.

He failed to mention that some of these were taken from Mantell's carefully researched illustrations and included Capt. Brickenden's famous Iguanodon jaw. He implied and would have them believe that all the illustrations were his own, stating that they were plates described by him in his 1841 report on 'British Fossil Reptiles'. At a special council meeting Mantell produced proof that some of the plates were not Owen's from 1841. Many of the fossils had not been found at that date. With this Owen was forced to backtrack. His anger knew no bounds and over the years he clashed with Lyell, Dr Alexander Melville, the biologist Hugh Falconer, Holmes, who had supplied him material from Tilgate forest, and the queen's dentist, Alexander Nasmyth, (who accused him of plagiarism), and eventually with Thomas Huxley.

In December, Mantell received a most welcome letter from Walter dated June, containing extracts from his 'Report to the Colonial Secretary on the physical structure of the Middle Island' (more usually referred to as the South Island) and numerous sketches. This interesting communication was backed up by a visit from Mr Alfred Wills who had travelled for several weeks with Walter in the Middle Island.

Despite fatigue and spasms in the chest (possibly angina), Mantell gave two lectures before the end of the year. At the Whittington Club in the Strand he spoke on the 'Colossal Birds of New Zealand' to a good and attentive audience which gratifyingly included Dr Fitton, and to a full audience on the 'Colossal Reptiles of the South East of England'. Somehow he lost two drawings of the humerus of Pelorosaurus he had displayed at the Whittington Club. To add to his misery he broke a tooth. Returning home in pain, at bedtime he used chloroform, fell asleep and remaining insensible with one of his feet in contact with a stone hot-water bottle. He woke in agony with the foot severely burnt and blistered. The lost diagrams eventually were returned having been found in an obscure corner of one of the rooms.

Reginald returned home safe from America after an absence of nine months. He had been treated with great kindness and befriended Prof. Silliman's son. He had received tremendous encouragement there and came away with the opinion that America posed a fine field for developments in railway engineering. So much so that Mantell very quickly realised that loneliness would be his inevitable destiny with the loss of both sons abroad. In February, Mantell demonstrated his specimens of Pelorosaurus and belemnites to the Royal Society where his papers were read to a very good attendance and the next day dined at the Geological Society. The following week he took his New Zealand fossils and Reginald's Oxford Clay fossils (from Reginald's paper on the Wiltshire, Somerset and Weymouth railway) to the Council of the Geological Society, where Prof. Owen caused a rumpus by declaring that the Moa supposedly had a hind toe to its foot. Then, at a meeting of the Zoological Council, Owen switched papers in order to present an unfinished one on the Dionornis. Mantell responded by bringing his Moa feet to the West of London Medical Society.

In May, Mantell took the opportunity to lecture at the Royal Institution on 'The Geology and Fossil Remains of the Birds of New Zealand', determined on Walter's account to keep the story of the colossal birds of New Zealand as much before the public as possible. His audience consisted of 700-800 people, a pleasing number even though he believed the institution to be aristocratic and attended in the main by fashionable lords and ladies. His friend, Professor Faraday, was most attentive, and insisted on supervising the hanging up of the drawings, and when the lecture was over, helped pack up the specimens, staying until eleven o'clock, and then saw Mantell to his carriage.

He was less happy that two papers sent to the Royal Society in November had not been printed and there were similar delays in the production of the *Pictorial Atlas of Organic Remains*. As always, when able to do so, he attended Professor Faraday's lecture and later on went with his son Reginald to soirées given by Lyell and Lord Rosse. Prince Albert gossiped with him about the Isle of Wight and thanked him for the book, asking him to send a copy of the *Pebble* for the Prince of Wales (Edward VII). He was later thanked by the Prince's tutor.

Mantell's next venture was a *Pictorial Atlas of Fossil Remains*, updating not only his own work but Parkinson's Organic Remains and Edmund Artis' Antediluvian Physiology. With the assistance of Henry Bohn and the younger John Morris (1816-1886), author of *A Catalogue of British Fossils*, he covered a range of Palaeontological topics from Fossil bears (quoting Buckland), sloths (citing Darwin), and saurians. He described the fossil reptiles, specifying how they had lived through the Mesozoic period starting in the Carboniferous and gradually declined at the end of the Cretaceous.

At the beginning of June he went with Reginald to the zoo to see a young hippopotamus which had just arrived and listened to two lectures by Faraday that week. Reginald had to leave for Reading, so Mantel went with Mr Alfred Woodhouse to Lord Rosse's last soirée taking their microscope. Among the throng of people present were the Rev. Richard Jeff DD, now Principal of King's College, London, who as tutor to the princes George of Cumberland and Cambridge had visited his museum in Lewes some fifteen years earlier. The Duke of Cambridge and his tutor examined the living floscularíae etc. on display.

On 29 June at about six o'clock Sir Robert Peel was thrown from his horse on Constitution Hill and picked up senseless with a fractured collar bone and severe concussion. Two days later, Mantell called at Sir Robert Peel's to be told that he was rather better; the crowd of anxious inquirers was very great. The accident proved fatal and he died on 2 July. It shook Mantell:

> This is a most deplorable event indeed; I am grieved exceedingly both on public and private grounds. For several years Sir Robert has shown me every courtesy; invited me to dine with distinguished foreigners and savants, apart from public policy. The last time I met this eminent statesman was at the Soiree of the Earl of Rosse, when he was in perfect health, and more affable and cheerful than usual; for his general bearing was cold and distant, and there were a stateliness and reserve in his demeanour, very repelling to those who were not intimately acquainted with him.

On 30 July he went with his son, Reginald Neville Mantell[1], to Euston Square and bade him farewell, in all probability forever. He sailed from Liverpool at noon the next day to work as a railway engineer at Paris, Kentucky. Mantell, a concerned father, refers to Paris as a violent frontier town, though it was probably at its heyday in importance. Thus, in 1828, it had featured a dinner attended by 4,500 in support of Cassius Clay in his dispute with Andrew Jackson. After several false starts the Lexington and Covington Railroad Company was granted a charter in 1849 and connected with the Mayville and Lexington railroad at Paris in 1854.

Fortunately Mantell was soon occupied with a trip north. On 1 August he left Euston travelling with Dr Wilson, an agreeable young American from Boston. At York he stayed with Professor Phillips and visited St Mary's Abbey and the Minster, continuing to Newcastle and Edinburgh the next day. With the sun setting Arthur's Seat rose up most majestically. On the Monday with 2,000 to 3,000 persons present and chaired by Sir David Brewster he lectured on the Geology of New Zealand; with a further lecture the next day on the dental organs of the Iguanodon and went with Mr and Mrs Lee to Roslyn chapel and castle. He returned via Kendal, taking a train to Windermere (then spelt Winandermere) crossing the lake on a steam boat and then to Manchester returning home after ten days.

Back in London he made an obligatory journey to Ticehurst. He left for Reigate by express train, met his brother there and after waiting an hour went on to Tunbridge Wells and took a fly for Ticehurst ten miles away; saw poor Joshua for half an hour; a sad painful interview (Joshua may well have failed to recognise him) and returned home where a specimen copy of *The Pictorial Atlas of Organic Remains* was ready for inspection.

In October he received by post two letters and a bundle of newspapers from Walter, brought across the oceans by the HMS *Woodstock*; and in the next week Walter's box arrived with bird skins of Apteryx (two species of Kiwis), Strigops, Neomorhs, Porphyrio and the Notornis – the only known specimen – from the southwest extremity of the South Island. After Mr Gould had made suitable drawings he tried to sell the Notornis for £25 to Mr Gray at the British Museum and had Mr Bartlett stuff and mount each bird. He was eventually to receive £25 for the specimen.

Despite severe sufferings with intense sciatica, made worse if anything from the inhalation of chloroform, he was able to visit Regent's Park with Miss Foster where the Grand Exhibition House was being erected; and lectured at the Whittington on 'Petrifactions and their Teachings' and in Salisbury on 'Corals and Coral Islands'. The year 1851 also heralded the death of several of his esteemed friends: the Marquess of Northampton died after a few days indisposition, Dr Pye Smith, Dr Basevi and Mrs Shelley (widow of Percy Bysshe Shelley and author of Frankenstein).

However, he was pleased to hear from Reginald in the USA that Professor Silliman and his family were due to visit Europe.

> They intended to sail for Liverpool about the 5th of March, and they pass right through our little island, calling at No. 19 Chester Square *en route* and sail for France, make their tour of the Continent and return to England and stay there and see the Exhibition. The old Professor you will be delighted with. He is the perfection of a true gentleman, and everything that is kind, good and affable. Dr B. S. jnr. is the best of men. I am sure you will like him as much as I do.

On Wednesday 26 March at half past seven in the evening,

> My beloved friend and correspondent and I met for the first time, after an interchange of thought and feeling of more than thirty years in my little study; he was accompanied by his son. We met as old familiar friends; he was what I had anticipated; a fine hearty intellectual countenance, and a frame upright and vigorous as a man of fifty – and I – alas! Non sum qualis eram! As poor Partridge says.

The Professor's energy at the age of seventy-two was wonderful. He was warmly welcomed by Mantell and at half past eight was formally introduced to the Geological Society, where he made a short address. Lyell, Murchison, De la Beche, Hopkins (who presided) and a tolerable number of old fellows were present. Prof. Silliman and his son put up at Morley Hotel. The next day Prof. Faraday called and gossiped awhile and brought tickets for a lecture on polarised light at the Royal Institution to which they all went with Professor Silliman and family. After the lecture they were introduced to a great number of the scientific community. The Sillimans dined with Mantell on Sunday the 30th and afterwards went to Westminster Abbey but Mantell was ill and remained on his sofa.

The next day the Silliman family headed for Paris where they were escorted to a meeting of the French Academy by Cordier, the sole survivor of the scientists who had gone to Egypt with Napoleon. They toured France, Italy and Sicily, climbing Vesuvius and exploring Etna and returned to Geneva where Professor Silliman met other Sillimani. His *Visit to Europe in 1851*, which he complied on his tour, achieved popularity going through several editions.

Though confined to the sofa for much of the time with severe neuralgia and fatigue, Mantell strived to keep himself active. He went with Mr Morris and Mr Alfred Woodhouse to the Royal Institution to hear a lecture by Lyell on 'Fossil Rain Drops', chaired by Prince Albert – a very crowded assemblage. Then later went to the new Museum of Practical Geology. Its very beautiful and interesting exhibition was officially opened the following month by Prince Albert with Mantell joining the very large assembly of scientific men and ladies.

Despite his poor health, on 1 May 1851 he walked to Constitution Hill to witness the Royal Procession going to the opening of the Grand Exhibition. He returned home and, suffering dreadfully lay on his sofa, but sent his servants to look at the buildings and the crowded scene. On the 9th he was fit enough to go with Mr and Mrs Allnutt to the Grand Exhibition:

> The coup d'oeil of the interior surpassed all my expectations. It is quite overpowering. I cannot express the effect it has left upon my mind. Stayed there four hours and returned home thoroughly exhausted.

The cost of visiting the Exhibition was £1 on days one and two, then 5 shillings per day and after 24 May one shilling per day. Mantell visited the exhibition many times, buying a season ticket at 3 guineas.

The Great Exhibition was the brainchild of Prince Albert. None of the original plans had pleased the Commission, but Joseph Paxton (the Duke of Devonshire's Head Gardener) who had constructed a magnificent glass house at Chatsworth submitted a design which was accepted. People thought that it would frighten the horses in Rotten Roy but the alternatives such as Battersea, Victoria Park and Kew were unsuitable. Rather than cut down the trees they were to be incorporated into the building itself. The structure was prefabricated off-site but the glass was hand-made and in order to test its strength 300 workmen, then a detachment of soldiers, and lastly boxes of loose shot were rolled upon it.

Queen Victoria was enchanted by the Great Exhibition:

> The glimpse of the transept through the iron gates, the waving palms, flowers, statues, myriads of people filling the galleries and seats around, with the flourish of trumpets as we entered, gave us a sensation which I can never forget and I felt much moved. The sight as we came to the middle where the steps and a chair were placed, with the beautiful crystal fountain just in front of it was magical – so vast, so glorious, so touching, One felt – as so many did whom I have spoken to – filled with devotion, more so than by any service I have ever heard. The tremendous cheers, the joy expressed in every face, the immensity of the building, the mixture of palms, flowers, trees, statues, fountains, the organ (with 200 instruments and 500 voices; which sounded like nothing) and my beloved husband, the author of this 'peace festival' which contained the industry of all the nations of the earth – all this was moving indeed, and it was and is a day to live for ever.

The Duke of Wellington commented:

> Whether the show will be of any use to any one may be questioned but of this I am certain, nothing can be more successful.

On a tightrope strung across the heights of Paxton's glasshouse, Blondin gave the Victorians every sensation until they were sick with pleasure. He traversed many times: in a sack, in an ape-suit and with his legs in shackles, cooking omelettes en route, and added a new routine with a suit of armour. He pushed a lion across the rope in a wheelbarrow as fireworks exploded overhead. He even crossed the rope carrying his five year old daughter Adele on his shoulders, until the Lord Chamberlain prohibited a repeat of this feat.

Mantell visited the Great Exhibition on at least nine other occasions spending on average three hours each time, braving the huge crowds, going or meeting up with friends, and examining Russian, French, Austrian and other displays. When the exhibition ended he was most anxious for its continuation in Regent's Park but eventually supported its transfer to Sydenham Hill Writing to Silliman while on the continent he could not resist comments about the Great Exhibition

> which engrosses everybody and everything. Lyell, Murchison, &c. are 'bowing at court and grown polite' and feasting and fete-ing at Lord

> Mayor's dinners and the Ministers' soirées, &c. I believe I am the only man of science who is plodding on his weary way – I am sorely delayed by the artists, lithographers and printers; all are engrossed in the Exhibition.

Despite his intense suffering, he continued a vigorous round of soirées, lectures, council and committee meetings. At the soirée of the President of the Royal Society, the Earl of Rosse, he showed bones and plants from the Wealden and palms from the Isle of Wight, gossiping over his fossils with the Bishop of Oxford (the Reverend Samuel Wilberforce, known as 'Soapy Sam') and examining other exhibits such as Wheatstone and Faraday's apparatus to show the motions of a vibrating wire when rotated in illustration of experiments demonstrating the rotation of the earth. At another soirée he showed Prince Albert the large scapula of Pelorosaurus. At the Royal College of Physicians he lent the President, Dr John Ayrton Paris, bones from a Pelorosaurus and chatted with Sir John Herschel on time, space and meteors. Less happily he attended a string of committee meetings at Somerset House where Owen was trying to obtain a government grant of nearly £1,000. As Professor Owen's demands became more and more strident, he felt heartily sick of repeatedly having to oppose the misapplication of money for Owen's use. Owen retaliated with misstatements in a monograph on cretaceous reptiles, by preventing Mr Bowerbank FRS from examining a specimen of a young Iguanodon, and launching a violent attack on Sir Charles Lyell's Anniversary Address in the *Quarterly Review* because Lyell had recommended a Mr Waterhouse to a post in the British Museum.

With his family dispersed and the need to be accompanied as often as possible because of his health, Mantell relied increasingly on his wife's nephew, Mr Alfred Woodhouse, a dentist and his frequent companion in the last four years of his life. His father was George Edward Woodhouse the younger. Unmentioned by Mantell in any of his journals or letters, his wife had visited him at Chester Square on 11 May 1850 for the first time since the funeral of Hannah Matilda, and was coldly received. He had had an uneasy relationship with Lupton Relfe, also related to the Woodhouses, who had been his publisher until he became bankrupt. Thus Mantell's continued bond with his in-laws appears remarkable. For example, Alfred Woodhouse accompanied him to Tunbridge Wells to look at the Wealden sandstone of the High Rocks, and to Calverley Park, to the infamous Thames tunnel and on to Southend and thence to Brighton. Mantell lectured to 300 members of the Medical Provincial Association of England in the Music room of the Royal Pavilion and on the next day, chaired by J. J. Marquerier, at a new institution got up by Mr Montagu Phillips called the Mantellian Academy of Science.

After four days on his return to Paris, Professor Silliman left the rest of the party and started for Boulogne. He found the steamer crowded with visitors for the Great Exhibition and the baggage examination took two hours. However, a gentleman, hearing his name and recognising him as a friend of Dr Mantell, ensured that his luggage passed unopened. On arrival in London he took a cab to Chester Square.

> I drove without hesitation, and was received with a very warm welcome, but with a kind reproof for lingering so long abroad, and leaving so little time for England. I found Dr Mantell uncommonly well and very spirited, and we entered at once upon a wide range of conversations.... The stream of his thoughts flowed on and on, rich, grand and inexhaustible... This part of London is new, clean, airy and quiet and this house is full of treasures of nature and art. I had enjoyed much on the Continent, but this was now a home and in it I was happy.

When Prof. Silliman returned from his trip in Europe they made an evening excursion on the Thames by a small steamer and, among other things, saw the Hungerford Suspension Bridge where Reginald had been working. Then they travelled together to the Isle of Wight on 20 August, returning to Portsmouth to see the American boat from the New York Yacht Club which was preparing to take part in the contest for the Prince Albert Cup. On the return journey Mantell took Silliman to his old haunts at Brighton and Lewes, and showed the professor a chalk-pit, a geological feature which he had never encountered. There were just a few days before Professor Silliman 'my excellent friend' left 'for ever! And so ends this dream'. The penultimate day was tinged with further regret, visiting the British Museum, they learnt that Mr Konig, the curator who had helped him in the purchase of the Mantellian Museum in Brighton, had died suddenly.

Mantell completed his work on the 'Petrifactions and their Teachings' and his handbook on the organic remains of the British Museum. There was a strong likelihood that Owen would be named to fill the British Museum post of Charles Konig. As Mantell wrote to Reginald:

> I have not spared him (Owen) in this new work (Petrifactions), and as this will be the first time anyone of my standing has reprobated his conduct, I expect the most vicious conduct from this unscrupulous man. He is now attempting to get the late Mr Konig's place in the British Museum and thus prevent Mr Waterhouse (O's supposed friend) from having the promotion he has been expecting. If Owen gets it (and I dare say he will), my collection and health will soon be deprived of an ally and I shall be

obliged to absent myself from the British Museum, as I have done from the College of Physicians.

The Telerpeton controversy (described earlier) was about to explode.

He went with Mrs Allnutt to the zoo to see a live *Apteryx manteli* just arrived from New Zealand. The appearance of the bird (a kiwi) was what his stuffed specimens and drawings had led him to expect. At the British Museum he learnt that Mr Waterhouse had just received the appointment. Prof. Owen had retired, and recommended Waterhouse! The Archbishop's letter the prior week to Mantell, convinced him that Owen's machinations would be defeated; which was indeed the case and he withdrew – 'Poor man: at all events my collection is safe from his clutches.'

It became obvious in 1852 that Dr Gideon Mantell's health was worsening and he was more and more frantic in his search to relieve the pains of his neuralgia and spasms. Nevertheless he attended the council of the Geological Society to sort out the argument with Professor Owen over the Telerpeton. Sir Charles Lyell was also involved, additionally quarrelling with Owen for having recommended Mr Waterhouse for the position at the British Museum.

More happily he met up with a Mr Evans of HMS *Acheron*, just returned from New Zealand, who gave a description of the country and sketches of the coast of the unexplored parts of the West coast of the Middle (South) island. The peaks of the gneiss and other primary rocks were from five to seven thousand feet high, intersected by vertical chasms extending for miles into the country and terminating with a bay or sound of deep water. Traditions of a race of wild men, and of Moas and Notornis, inhabitants of the unexplored recesses, are rife among the natives and whalers. Mr Evans and party killed thirty-forty Apteryxes and many Strigops. Walter's Notornis was believed to have been driven by stress of weather from the heights to the coast at Dusky Bay. He was also glad to receive a copy of Mr Dixon's work on the Sussex fossils from his widow, through the kind suggestion of Mr Morris.

He went with Mr Woodhouse to the Egyptian Hall, Piccadilly to hear a lecture on electro-biology (in effect on mesmerism), which he thought offended the principles of physiology. At a party on the anniversary of Mr Allnutt's birthday – his eightieth, he described the birds of New Zealand with diagrams and exhibited the feet and bones of a Moa. In March he gave a lecture at the Royal Institution. Mr Beckles had given him an interesting assortment of fossils including (he believed) the radius and ulna of the Iguanodon; the first he had ever seen, as well as a humerus probably belonging to a new species of Pelorosaurus. After an hour and twenty minutes discourse to a crowded lecture room he had to stand at the table for a further half hour to explain the diagrams and fossils. Among those present that day were Sir C. Lyell, Capt. Brickenden, Mr Grantham and Prof. Faraday.

After a night of agony on 13 March with no relief from prussic acid, fomentations, liniments, calomel and opium, the paroxysms became less severe through the morning and he dressed with difficulty for the soirée of the Duchess of Northumberland at Northumberland House. Mr Alfred Woodhouse with great kindness accompanied him to the house and waited in his carriage. Prof. Faraday helped him upstairs where, in a suffocating atmosphere he gossiped with Sir John Herschel and others. He compared the liveliness and happy humour of the Duke of Wellington, who chatted in a corner of the room, with Sir R. Inglis whom he saw at the Royal Society the same evening looking as lethargic and automatic as the effigies of some of the old worthies that decorated the walls.

Ill health did not reduce his capacity for comment. When he attended the soirée of Sir R. Murchison, President of the Geographical Society, in yet another crowded room, he chatted to Robert Chambers, the author of *Vestiges*. His comments on the work were more mellow, but he castigated the elite of the fashionable world and scores of hangers-on of scientific societies, who ought not to have been admitted to such a circle. During 1852, complaining that his memory was failing dreadfully and that he had to rely on notes, he worked on a second edition of *Medals of Creation* hoping that it would be a financial success.

In April he was busy with a series of lectures but even so found time to revisit the Crystal Palace where some 70,000 people were present. The crowd was as great and as orderly as in the previous year. He gave an address as President to the West London Medical Society before leaving with his footman, John Canner, from King's Cross to Leeds, visiting friends and interesting sights in the neighbourhood and giving four lectures at the Mechanics Institution. On his return, he went to the Westminster meeting to add his support in petitioning Parliament for the perpetuation of the Crystal Palace and later on went to the Exeter Hall for the same purpose. Nonetheless on 1 May the decision was taken to pull down the Crystal Palace though in the face of objections it was later decided to reconstruct it at Sydenham Hill. The one evening meeting that enthralled him was at Lord Londesborough's with tables of antiquities to which he brought the glass Roman cinerary vase from Puttioli, given him by Sir Woodbine Parish.

Virtually the last entry in the journal which went to Walter in New Zealand was a soirée at Lord Rosse's with Mr Alfred Woodhouse. Arriving early, he went with Lady Rosse and children round the rooms and looked at the various and beautiful objects of science and art that had been sent for exhibition. The drawings of the Moa (ten foot high) were suspended over the crimson curtains of one of the drawing room windows and looked very imposing with the bones placed on a nearby table. All the specimens were returned to safety the following morning.

The journal rescued by Reginald runs from June to November. The state in the person of the Minister, Lord Derby, offered an annuity of £100 to Mantell, as an expression of respect from the crown for his scientific labours. While on the one hand he realised that this was a derisory amount; he nonetheless thanked Lord Rosse who made the application on his own accord as 'I am so little known to his lordship personally that it is the more surprising to me. It is very gratifying in every sense'. He called on Lord Rosse to thank him. Though when in October he called at the Treasury to obtain the first quarter's pension, it was reduced to £24 4*s* 5*d* after deducting income tax and reduced still further to £21 for the fee of £4 stamp of warrant!

His daughter Ellen Maria called with her boy after an absence of nine years. He was happy to receive her and they dined together. Later he gave her a pair of bronze chandeliers, and as late as November some table cloths. He was clearly very ill during this period but strove to be active; however on 27 July, returning home over Westminster Bridge his coachman got the carriage entangled between two waggons in the middle of a dangerous and scandalous pass, and got away with only the carriage step twisted off but a narrow escape of 'my life'. He developed a series of generalised symptoms in October. In addition to the neuralgia he complained of face and neck ache and dyspepsia, and was obviously an ill man.

Despite his parlous state of health, among numerous activities, he went with Mr Alfred Woodhouse to Brighton then to Hastings and on to Battle to attend a meeting of the Sussex Archaeological Society with 300-400 Sussex gentlemen; and later to the opening of the new Crystal Palace at Penge Hill, Sydenham along with a crowd of 10,000. He presided and addressed the opening meeting of the West London Medical Society on 15 October and despite persistent illness kept himself busy correcting proofs of *Medals*. On 5 November he attended the council and general meetings of the Geological Society. Less than a week before he died on 10 November he attended several patients, including Mrs Kemble of Stockwell; but he felt very unwell with neuralgia 'flying from one limb to another' and great irritation of the abdominal viscera.

On 8 November he gave a lecture for more than an hour at the Clapham Athenaeum., being now and again carried away by his subject despite his morbid state with frequent 'nervous shocks' and the need to be seated for most of the time. Immediately afterwards he complained of extreme exhaustion and had to be rushed home. He spent all the next day doubled up in constant pain. He told his nephew, Alfred Woodhouse, that he must have an opiate if he could not sleep but hoped to avoid it as it always made him so ill afterwards. He retired at the extremely uncharacteristic early hour of 8.00 p.m., falling on his way, slipping on the stairs, and obliged to crawl on his hands and knees until he reached his bed. He then took half the draught which he had sent for, but that not having the desired effect, he finished the remainder at 10.00 p.m. At 12.30 he rang for his servant to sit with him until he went to sleep. The next morning he was comatose and at around noon Mr Woodhouse was sent for. Consciousness was not regained and he died at 3.00 p.m.

A year earlier Mantell had placed an envelope with Alfred Woodhouse to be opened on his death. In it he asked that a post-mortem examination be made of his spine and right elbow in the belief that the enormous changes which have taken place may be interesting in a pathological point of view; if any relevant parts are worth preserving for examples of morbid changes they should be given to the Hunterian Museum of the Royal College of Surgeons. P. J. Martin relates that

> I jested with him about leaving of the honour of the extraordinary tumour of his spine to Owen. In the same humour, he replied 'No, No! Owen shall not have the pickings of my bones! I mean to depute the task to my friend Hodgkin.'

Other instructions were that the funeral should be as plain as possible, should take palace in the early morning at Norwood cemetery and he should be buried next to his daughter Hannah. A few selected mourners came to the ceremony. He had drawn the outline of the plain block monument he wished to place over the two graves and had asked Reginald to place a plaque of Sussex marble holding a brass with an ornamental border, a coat of arms and a long inscription in St Michael's church, Lewes below that to his father. When alterations were later made to the church the plaque and the tablet to Thomas Mantell were repositioned.

In his will, with Alfred Woodhouse and Stephen Williams as executors, he left annuities of £50 to his wife, £30 to Mrs Best and £15 to Kezia. Catherine Foster received the rosewood table with marble septatia top and the miniatures of Napoleon, Silliman the five volumes of Cuvier's Ossemans Fossiles and the other possessions to Reginald. £10 was set aside for Dr Hodgkin to perform the post mortem. Three days after his death two obituaries appeared:

That in the *Sussex Express* was laudatory, they had interviewed him a month earlier:

> his mental powers were not only unimpaired but as lively as ever. However even as he spoke we felt that the unerring, though silent, hand of death was laid upon him. We feel in his death that we have lost no ordinary man.

The other, anonymously written in the *Literary Gazette*, was full of misinformation and snide deprecations.

> He unfortunately lacked exact scientific (and especially anatomical) knowledge, which compelled him privately to have recourse to those possessing it. In his popular summaries of geological facts he was too apt to forget the sources of information which he had acknowledged in his original memoirs. The history of the Iguanodon provided a remarkable instance of this. To Cuvier we owe the first recognition of its reptilian character, to Clift the first perception of the resemblance of its teeth to those of the iguana, to Conybeare its name, and to Owen its true affinities among reptiles, and the correction of errors respecting its bulk and alleged horn.

Owen's mean-spirited jibes in this obituary cost him the Presidency of the Geological Society of London.

HIS PERSONALITY AND HEALTH

From his earliest days, Gideon was considered by his family to be exceptionally bright with a retentive memory. He blossomed in Button's Dissident Academy, receiving as modern and broad an education as was possible at that time, with encouragement to write poetry, draw and speak in public, achieving fluency with clear articulation and expunging 'vulgar and provincial' diction. Then, as an adolescent, he was sent to Wiltshire to be instructed under the guidance of his uncle, Rev. George Mantell, a Congregational Minister who had bridged the social gap between trade and profession. As far as is known, all of his siblings spent their school days within Lewes. However, he was concerned that the comparative poverty of his origins and professional status would set him apart from his wealthier contemporaries unburdened by any need to earn their livelihood.

His mother possessed a certain dynamism and was regarded as the ruling spirit of the household. If she was the dominant person in the household, why was it necessary for Gideon to be fostered? There is no way to ascertain why this was so but one can speculate. The house in the steeply sloping St Mary Street was probably quite small and congested. In all likelihood she had little or no additional help. Gideon was the fifth child, one dying shortly after birth, and his mother had three more children, the first being Joshua who was delicate at birth and thought unlikely to survive. Thus, it is likely that, at the time of childbirth, Gideon was hastily fostered out as an infant to a lady nearby. Mrs C was almost certainly not a relative. She may have worked for Mrs Mantell at an earlier date, or she might have been a neighbour and a member of the congregation of the chapel. 'Mrs' may have been a courtesy title. She could have been a spinster or a widow and it is possible that she derived an income from fostering children. This supposition gains credence when, slightly older, Gideon was sent not to Mrs C in Lewes but to her brother, Mr Cornwell, fifteen miles away near

Cross-in-Hand where he spent several weeks. Revisiting the cottage in 1820, everything seemed to be on a smaller scale than his recollection had painted.

His father, an artisan employing many journeymen and apprentices, has been described as unlettered but read voraciously with a deep involvement in religion and Whig politics. Through his interest in such matters he established a strong friendship with the Lee family who published the first newspaper in Lewes. Respected for his shrewdness and integrity, he was an excellent businessman, and, as is evident from his will, owned numerous properties. As a pillar of the community, he built the first Wesleyan chapel in town. Likewise, Gideon's eldest brother, Thomas, sharing the civic and work ethic of his father, was a Sheriff's officer and auctioneer, taking over primary responsibility for the family when his father died in 1807. The second son, Samuel, probably had no claim to academic ability and became first a butcher and later a publican. The younger brother, Joshua, was delicate at birth, which may have affected his education, but was bright with a scientific bent. One sister died in infancy; two others married and little is known about their abilities. The fourth, Kezia, was employed as a servant and may have lacked intelligence.

From the beginning, the atmosphere in which Gideon was brought up was earnest with serious debate on religious and political matters, and the standing of dissidents within a class structured community was allied to a need to help one's fellow man. Being by nature a hardworking, gifted child with abundant restless energy, his inquisitive and acquisitive temperament led to a fierce desire for self-education. He continued this disciplined existence thoroughout his life, devoting many hours to study, using lending libraries, seeking news of developments and advances in science, walking the Downs, studying the strata, examining the flora and fauna of the vicinity and befriending people such as the Rev. James Douglas, bent on investigating and collecting the antiquities of the past. The people of Lewes, especially the dissident community, were still influenced by the republicanism of Tom Paine. They despised the class structure of society and were ambiguous and equivocal with respect to the French Revolution. There was also an element of an inferiority complex, a desire to burst out from humble origins, fortifying an ambition to become immortal in the realms of science. Thomas Arnold (1795-1842), headmaster of Rugby School, claimed that he could take the son of a successful tradesman and in three generations create a gentleman. Gideon Mantell was in a far greater hurry.

The process of self-education continued apace throughout his apprenticeship and training as a surgeon, constrained by a strong discipline and financial limitations. In early life his health appeared to be robust, enabling him to work long hours and achieve more than was expected of him. Once qualified he was determined to achieve financial viability and, by hard work and long hours, establish an impressive, if ostentatious, social presence. He was not only conscientious as a doctor, with considerable empathy for his patients, but anxious to improve his skills and advance medical understanding. In many respects this was the background to his scientific endeavours on which he worked late into the night, also writing his journal, corresponding with fellow scientists and others, and existing on only a few hours of sleep.

In March 1816, Gideon Mantell moved into Castle Place and two months later married Mary Ann Woodhouse. He was twenty-six; she, being just twenty-one, required the consent of her mother. 1816 could be called an inauspicious year, the 'year without a summer', due to the change of climate brought about by the eruption of Mt Tamboro in the Dutch East Indies in 1815. His wife shared his enthusiasm for fossil collecting, astronomy, the theatre and chess, and he warmly acknowledged her contributions illustrating his *Fossils of the South Downs*, making landscape drawings using a camera lucida, and supposedly finding the first Iguanodon tooth. He had many reasons to be grateful to his wife's family, with whom he remained on good terms. George Edward Woodhouse Snr had presented him with all three volumes of James Parkinson's *Organic Remains of a Former World*; Lupton Relfe, married to his sister-in-law, was initially his publisher and George Edward Woodhouse Jnr, a lawyer, defrayed the costs of the publication.

Gideon Mantell was devoted to his family, taking them with him whenever possible and administering to their physical illnesses. Early on, his wife suffered from asthma and a bloody expectoration, his elder daughter from a bowel affliction and his elder son from a chest complaint. There was no lack of closeness or support, but, arising from his rigid personal astringency, he became a domineering martinet, rendering it difficult for any family member or servant to assist him in a subordinate role. His thrusting ambition caused him to fill his house with opulent artefacts and accumulatively with what became his museum collection. Increasingly, he was unable to empathise with the mental worries and traumas of others. For example, Mary Ann became weakened and depressed after her first, unbaptised child died a few days after birth in April 1817, and she was later to suffer from further bouts of depression, many of which were consequent on the cramping of the domestic arrangements of the household.

His rigid domination of those around him left him without anyone with whom he could exteriorise his doubts, his worries, or difficulties with some of the personalities among his fellow scientists. His geological pursuits meant that he exchanged boxes of fossils with others and to

Rouse Keep) The remains of the keep of Lewes Castle, drawn and published by James Rouse, Fulham, 1833.

some extent gained their friendship. He also corresponded with the elite of his day, Herschel, Dickens, Lytton and others; but he needed more regular correspondents who would help him through his introspection and minimise his paranoia; for there were times when he felt a coldness even to his supporters such as Buckland, Murchison and Lyell. Over the years he developed a more intimate form of correspondence with Samuel Woodward of Norfolk, author of an *Outline of the Geology of Norfolk*; with Professor Benjamin Silliman in America whom he did not meet until the penultimate period of his life; and with George Grantham of Barcombe, thus retaining links with Lewes.

His letters and correspondence frequently reflect changes in mood, certainly when in Brighton he had cause to be depressed as his medical practice fell apart and family quarrels became more and more persistent. He also had periods of elation. However, though his activity may have been manic at times, he does not fit into the recognised pattern of a manic-depressive with changes in weight and sleep patterns, nor could the term 'rapid bipolar' be strictly applied. His mood changes could be accepted as normal or perhaps as that of a cyclothymic personality with a lifelong tendency to fluctuating depressive and elated moods. In Lewes, his demanding practice, part town, part country, meant contact with unsanitary conditions and the prevalent fevers such as typhus in 1818, typhoid and cholera. He made his rounds in all weathers, working sixteen hours a day and travelling by carriage or on horseback, but often obliged to walk if the carriage broke down or a horse became ill. Accidents were not uncommon. In 1819 his horse was struck by lightning in a thunder storm. In December the following year, his horse fell near Ranscombe owing to the badness of the roads and he escaped uninjured; but in March 1830 he was thrown from his horse, bruising his shoulder and necessitating the application of leeches.

On another occasion, in December 1826, his horse started suddenly as he mounted, resulting in a right inguinal hernia for which he subsequently wore a truss. There is some confusion, through lack of anatomical detail, whether there was a complication of the hernia or whether a separate tumour developed the next year in his groin to which he applied leeches and later an unusual preparation in order to prevent suppuration. Such a tumour could have arisen from chaffing or contact with animals; a venereal cause is unlikely, but he later suffered from a bladder infection and worried that it might indicate disease of the prostate gland. He also complained of lumbago which could be explained by a very exhausting excursion over several days. Incidental illnesses were often quite serious. He applied fifteen leeches to his head after a bout of giddiness and had a series of chest infections. In December 1827, when his wife was also ill, he was indisposed for several weeks but recovered amid much anxiety and harassing duties.

In 1830 he complained of rheumatism after exertion, suggesting the possibility of angina, and in November of that year was ill in bed all day and was then woken from a deep sleep, induced by taking strong opiates, to the sound of Bellman's bell and cries that a barn had been set on fire. Yet again that month he was ill with a severe headache, fever and disturbance of vision. 'Kept to my bed all day; last night became quite delirious and for several hours after I went to bed my head was in so confused a state that I made a vain attempt to recall my wanderings.' What is apparent is that towards the end of his time in Lewes he had a breakdown in health sufficiently severe for him to need to take on two partners to his practice. He later told his correspondents that he had been ill but was now convalescent.

During that decade he had numerous minor illnesses such as toothache, headache, and in May 1835, after sitting up late at night on Brighton's Chain Pier with a guest, had a severe attack of internal inflammation. What that amounted to is unclear except that he then complained of frequent attacks of lassitude and general malaise.

His failure to establish himself as a medical practitioner in Brighton resulted in his becoming seriously depressed and physically rundown. As he wrote to Professor Silliman, 'My health has cruelly suffered, so that from being a man with a very youthful appearance for my age, I am becoming more worn than my years would warrant.' Somehow, despite his depressed state, he was able to work on his lectures, culminating in his most popular work, the *Medals of Creation* (1838). In September 1835 he developed a swelling of the face and two days later the abscess in his cheek burst, easing his pain. The following February he complained most dreadfully of tic douloureux. It is possible that this diagnosis was not accurate, however.

Mantell states that:

> The most unfortunate thing that has occurred to me is an attack of tic douloureux, or neuralgia, as it ought to be called, affecting one side of the face, coming on in severe paroxysms and not being even alleviated by any means I have hitherto tried. I attribute it partly to mental anxiety and partly to the extraordinary wet weather accompanied with violent winds, and Brighton is more exposed than Lewes. Even now I write under a most severe paroxysm, trying to keep my mind employed in the most grateful and pleasing of occupations and thus be less the sport of physical influences – but alas! It is almost in vain, and I must for a while lay down my pen!

Considering Mantell's description of the neuralgia, his diagnosis becomes questionable. Tic douloureux, trigeminal neuralgia, produces recurrent, intense lancinating pains often accompanied by fleeting contractions of the facial muscles. Attacks often last several weeks with a history of recurrence after periods of remission. Occurring mainly in the elderly, the general trend is one of progression in frequency, intensity and extent. A spontaneous recovery or cure is a rarity. What renders the diagnosis most unlikely is that he fails to refer to trigger factors which induce paroxysms such as touching the face, talking, washing, shaving, cleaning the teeth, and eating.

Atypical facial pains can lead to diagnostic difficulty. The pain is often deep, unilateral, lasting hours or days and occurs, generally, in a younger age group. It may be associated with other factors such as temporo-mandibular joint dysfunction. In my experience, not all atypical facial pain is psychogenic but it is often regarded as such and can be associated with manifestations of chronic anxiety and depression. Facial palsy (Bell's palsy) can be painful, especially if associated with a herpetic viral infection.

A third diagnosis would certainly fit the circumstances, namely periodic migrainous neuralgia (cluster headache), where attacks of severe pain in or around the eye last up to two hours and occur daily or twice daily in bouts lasting for several weeks. It is distinguished from trigeminal neuralgia by its periodicity, the absence of precipitating factors and the much longer duration of each paroxysm.

In July 1839 he complained of severe fainting fits and suffered excruciatingly from tic douloureux and all the attendants of a highly excitable state of the nervous system. In September 1840 there was no further mention of the painful and excruciating tic. Thus the most probable diagnosis is of psychogenic atypical facial pain. After the death of Hannah Matilda, he was undoubtedly depressed but his health improved in September 1840.

The accident of 11 October 1841 has been depicted as the irretrievable turning point of Mantell's life; from then on his health was forfeit, a downward descent to a painful death. Mantell described it to his friend as the culmination of a series of back problems resulting from bending over low beds to administer to his dying daughter and to various other patients requiring his help. There is no certain way to gauge the severity of the accident. Did the wheel of the carriage do more than graze his head? Did he fall awkwardly on his back damaging the lower part of his spine? Having fallen, was he dragged along? We know from the post-mortem examination of his spine eleven years later that it was twisted and distorted with a lateral curvature, though the spinous processes remained in alignment. In the intermediate years he had been examined by the doyen of the surgical and medical professions who disagreed as to the exact nature of the tumour that came, appeared to fluctuate, almost disappeared, and returned larger and harder.

The origin of his back problem almost certainly dates back to an asymptomatic, adolescent, idiopathic curvature (scoliosis) tending to progress slowly over the years as suggested by the orthopaedic surgeon, Jeremy C. T. Fairbank (2004). Dr E. C. Curwen (1940), who had examined the plaster cast and specimen in the Hunterian collection, postulated an early stage of pain and discomfort from osteo-arthritis and postural scoliosis, unnoticed for many years. In fact, when aged thirty-six (20 June 1826), Mantell complained of intense lumbago after returning from a journey of several days covering 375 miles, and in July 1830 was unwell from rheumatism after the previous day's exertion. Fairbank also drew attention to the Adams test, named after William Adams FRCS the orthopaedic surgeon who performed the post mortem with Dr Hodgkin on 13 November 1852. Adams showed that bending forward could confirm and exacerbate a lumbar scoliosis. In a further comment Fairbank stated that lumbar curvature can give rise to back pain which is often of the fatigue type where the pain worsens through the day and may be enough to make the subject lie down by the afternoon or evening. Mantell seemed to have suffered particularly from pain at night.

From the autopsy we learn that, despite his cachetic appearance, he was a tall, well-built, muscular man. The body was examined from the back. By palpation, three or four hard nodules could be felt situated two to three inches to the left of the spinous processes. These were the transverse processes of the vertebrae projecting backwards and rising to the same level as the spinous processes. The abdominal cavity was opened and examined; and finally the lower three dorsal vertebrae, all the lumbar, and a portion of the sacrum, were removed from the back. There was a very severe lateral curve to the left involving the vertebral bodies and transverse processes. The aorta appeared normal. And there was no thickening, adhesions or other alterations to any of the soft tissues. The musculature, ligaments and cellular tissue were healthy with no trace of an abscess, malignancy or inflammatory process. The transverse process of the fourth lumbar vertebra was remarkably altered both in direction and form, curving upwards, close to that of the third vertebra, compressed from above downwards and expanded horizontally to articulate with the ilium. The lateral concavity of the curve arose mainly from absorption of the inter-vertebral cartilages, especially between the bodies of one-two and two-three lumbar vertebrae. Minor alterations of the osseous structure indicated a reparative process with considerable increase in the thickness and density of the outer surface of the bodies of the vertebrae, and the adjacent portion of cancellous bone.

What exactly happened is a conundrum. If we follow Dean's account: after walking home on an intensely cold night, pains already in his lower back intensified, and his left foot numbed, with paralysis of his legs and pelvis following. He remained at home for the rest of the month, continuing to experience numbness and immobility. Elsewhere we are told that on 17 October he was ill with symptoms of paralysis arising from spinal disease and on 17 November, in writing to Professor Silliman, he says that for the last three months he had been suffering from a paralytic affection. What is clear is that following the accident he developed an ascending paralysis even affecting his right arm, with loss of sensation of limbs and pelvic viscera; so that he required a catheter for his bladder and enemata to relieve his bowels. With regard to the pain we do not know whether it was the long-standing pain, a pain arising shortly before the accident, as some reports suggest, or back pain due to the accident. He might have had a slight temperature. The timescale remains uncertain. Fairbank makes two suggestions how the symptoms may have arisen: (a) although a cauda equina syndrome is not a normal concomitant of scoliosis it might be precipitated by a disc prolapse; and (b) retention may be precipitated by severe pain, especially if treated with large doses of opiates. Neither of these suggestions may be totally dismissed out of hand; but there is no evidence that Mantell resorted to his usual means of relieving pain by either leeches or heavy doses of analgesics.

Fairbank regretted that it is not possible to examine Mantell's spine by CT or MRI scanning. The specimen was supposed to have been destroyed by German bombs in the Second World War during the Blitz but Anthony Brook (2007) has established that the plaster cast was discarded in 1945 and the actual specimen survived until 1970 when it was discarded along with other poorly documented materials. Unfortunately at the time of dissection it was not possible to examine the vascular supply to the lumbar region, and Hodgkin and Adams chose not to open up the spine to examine the nerves of the cauda equina. In dissecting out the three lower thoracic vertebrae they would have seen the lower end of the spinal cord and beginnings of the cauda equina and they could have examined the exit foramina and nerve roots, but no record was made. In addition to Fairbank's suggestions, the sequence of events might just be explicable by an acute transverse myelitis or a vascular event leading to an ischaemic myelopathy such as hemisection of the cord (the Brown-Sequard syndrome).

However, in my opinion the most probable cause of the ascending paralysis with loss of sensation affecting the limbs and pelvic viscera is a Guillain-Barre syndrome (acute idiopathic or acute infective neuropathy) precipitated by a hypersensitivity reaction or a mild infection. Pain and a slight temperature may occur in the early stages; the symptoms tend to fluctuate and in the milder cases start to improve after about six weeks. More persistent changes may take months before improvement occurs but the impairment they produce may fail to be corrected. The musculature would be weakened in the acute stage with wasting and loss of tone in anything but a uniform manner. Any existing scoliosis would lose stability and most certainly rotation of the vertebral bodies could occur. Emerging nerves would be squeezed as they emerge from the exit foramina of the spinal column; and this could vary as the degree of rotation progressed. It may prove difficult to determine what has been the direct result of the initial illness and the subsequent changes related to the instability of the spine. An acute neuropathy of this nature is the only diagnosis which can account for the involvement of the right arm. Cardiac irregularities can also occur.[1] One persisting effect of the syndrome would be a weakened musculature allowing the scoliosis to progress and give rise to later symptoms of nerve compression with neuralgia and spasms. The patient would continue to have good and bad days.

In 1846 Professor Silliman asked Mantell for further details about his condition. In his reply on 29 May he omitted all reference to the accident:

> Soon after the death of my dear Hannah, I found my back very bad; but leeches relieved me, and had I then rested a few weeks I should have warded off more serious evil; but this unfortunately I did not do. In the following October, one bitter cold night, I had been in attendance for hours over a lady who had fallen down stairs and was apparently dying from concussion of the brain. At four in the morning she was sufficiently recovered for me to leave her, and I went out of the hot apartment into the cold air and walked home (about half a mile) and went into a cold bed; which I felt at the time exceedingly. When I awoke the left thigh and leg were completely paralysed and the right partially so – but you know the rest. After very slow amendment as to the palsy, severe neuralgia (a frequent sequel) came on. At length a small tumour appeared on the left side of the lumbar vertebrae (externally) which gradually increased: it was very hard, the nature doubtful.

If Mantell's recollection was correct, nearly five years after the incident he appeared to have had Landry's ascending paralysis, a form of the Guillain-Barre syndrome.

From the time of the accident we can have no doubt about the physical nature of the majority of his symptoms. His fortitude in relation to the pains, spasms and neuralgia was impressive, though nothing like as remarkable as his activity and enormous output of books and lectures during that period. The decade after his accident saw a remarkable flourishing of research, particularly that which involved microscopy. Perhaps his anger with his detractors such as Professor Owen did much to mitigate the effects of any depression he may have had. From the ultimate part of his journal, rescued by Reginald, there is evidence of a very definite deterioration in his health and appearance in the last few months, and he complained that his memory, which had been remarkable, as evidenced by his knowledge of poetry, was failing him so that he had to rely on notes. His neuralgic pains were erratic and excessive: shifting, unpredictable and widespread. He became increasingly dependent on drugs; but although the suggestion of suicide has been made, his demise was inevitable rather than deliberate (see Brook).

To end, like his spine, with a morbid twist, Spokes describes 'A Dinner in the Iguanodon' on New Year's Eve 1853 after Mantell's death, in the park by the newly erected Crystal Palace at Sydenham. The Directors, seeking to devise a great educational scheme, had entrusted Mr Waterhouse Hawkins to produce lifelike restorations of several of the 'Extinct Inhabitants of the Ancient World'. Dr Mantell was the official adviser, though he never saw the completed exhibit; and on his death Richard Owen took over the task. The banquet was held inside the concrete carcass of the standing

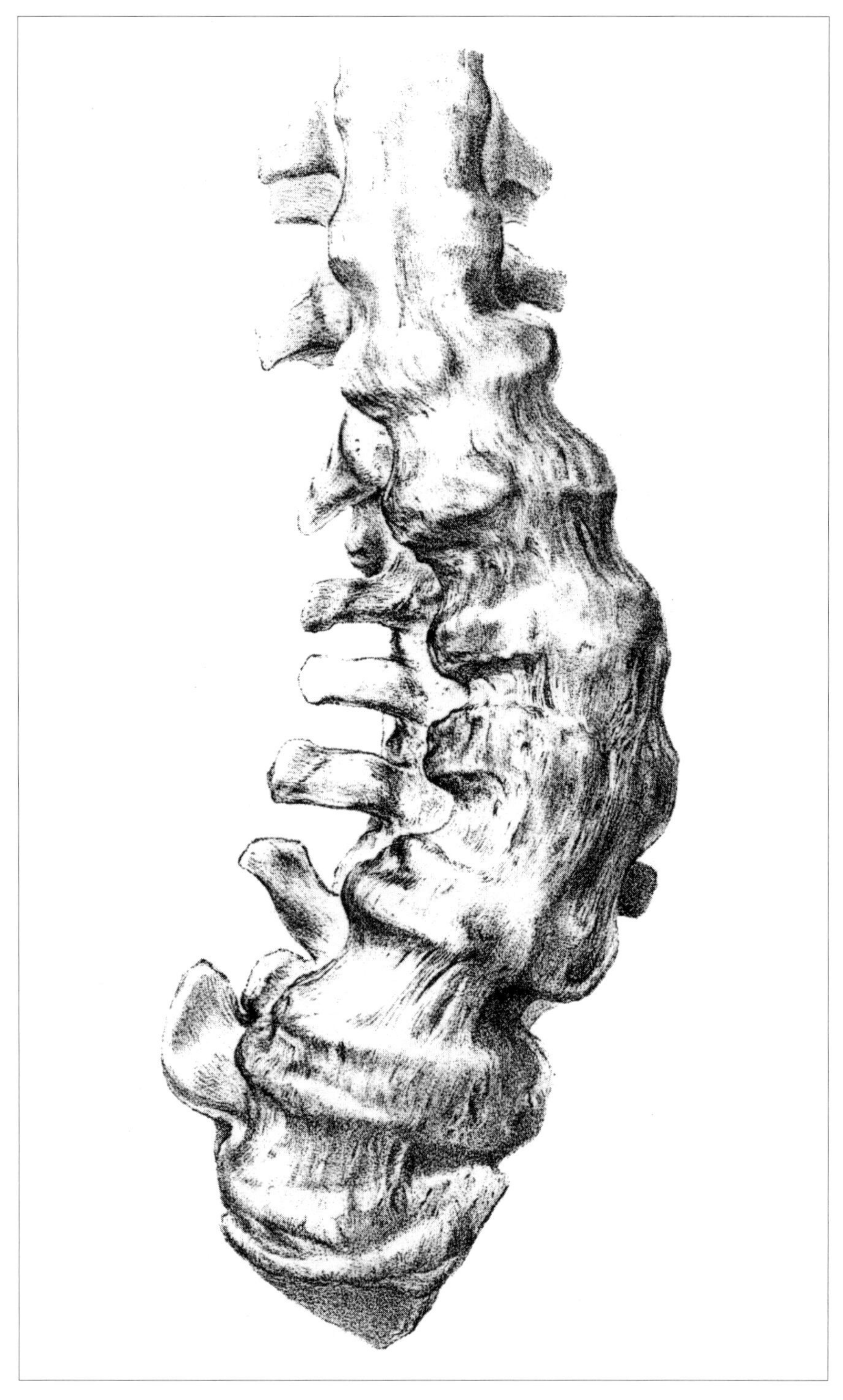

Above and opposite: Mantell's spine From the Hunterian Museum of the Royal College of Surgeons, London.

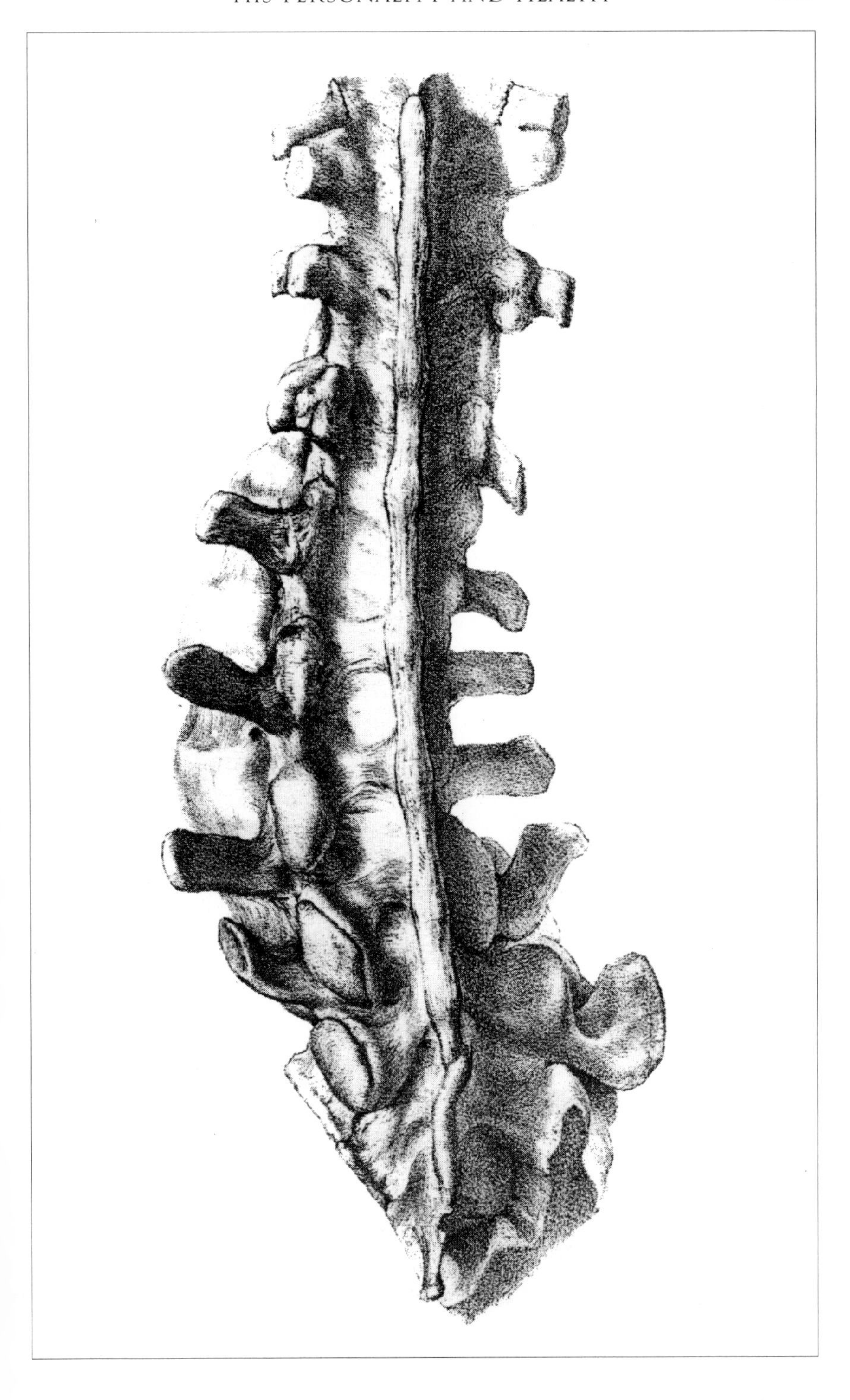

Iguanodon beneath awnings of pink and white with draperies and banners to Conybeare, Forbes, etc. The invitations were written on the wing of a Pterodactyl and, at four o'clock on the last day of the year, twenty-one guests sat down at the feast. This was also shown in the *Illustrated London News*. On the right side of the animal there was apparently an extension, at the end of which sat the host, but by the head of the creature Owen is shown, surrounded by palaeontologists and geologists. On the walls of the saurian appear the names of Mantell, Owen, Cuvier and Buckland. Of these Owen, the only survivor, is represented raising his glass as if to salute the name of Mantell, which happened to be suspended opposite to him. Speeches followed. Professor Owen took the occasion to explain, in his lucid and powerful manner, the means and careful study by which the models were prepared and attained a truthful success. It is interesting to note that the illustration shows the 'horn of the Iguanodon' well displayed on the animal's head so that Owen, in his testimony as to the truth of the restoration under his special guidance, commits an error, for in reality the feature was a claw and not a horn. The banquet ended at midnight when Mr Hawkins and his scientific friends broke into song:

For monsters wise our Saurians are
And wisely shall they reign
To speed sound knowledge near and far
They've come to life again.

GLOSSARY

HORSE-DRAWN TRANSPORT

AGE	Anything one pays for, presumably an abbreviation of Stage
BAROUCHE	4-wheeled carriage with a retractable hood over the rear half, inside couples face each other with a driver's seat on the outside.
BERLIN	4-wheeled, two seated carriage
BROUGHAM	4-wheeled with a raised driver's seat
CABRIOLET	a small 2-wheeled carriage with 2 seats and a folding hood
CALASH (CALACHE)	a carriage with low wheels and a folding top (or its folding top)
CARRIAGE	4-wheeled horse drawn vehicle
CHAISE	2-wheeled light carriage designed for 2 passangers
CHARIOT	any stately vehicle
CLARENCE	a closed 2-wheeled carriage having a glass front
COACH	4-wheeled enclosed carriage for passengers over distance
CURRICLE	2-wheeled open carriage dra\wn by two horses side by side
DILIGENCE	a stage coach, from the French carose de diligence (coach of speed)
FLY	a light one horse covered carriage , formerly let out of hire
GIG	a light 2-wheeled carriage without a hood
LANDAU	4-wheeled carriage with 2 folding hoods which meet over the middle passenger compartment
PHAETON	a light 4-wheeled carriage usually with 2 seats, with or without a top
SULKY	2-wheeled vehicle for one person drawn by one horse
SURREY	a light 4-wheeled carriage with 2-4 seats
TRAP	a light 2-wheeled carriage
VICTORIA	a light 4-wheeled carriage with a folding hood, 2 passenger seats and a seat in front for the driver

ENDNOTES

INFLUENCES

1. After losing the Battle of Worcester and hiding in an oak tree, the uncrowned Charles II arrived in Sussex in disguise with a price of £100,000 on his head for the 'tall black man, over 2 yards high'. He travelled through Sussex with the help of Col. Gunter of Racton, arriving at Brightelmstone and possibly (according to Horsfield) was sheltered by Mr Maunsell at Ovingdean, before escaping to France from Shoreham. A quatrain by Charles Dalmon described how he looked wistfully at Chanctonbury circle:

    The king was making for the shore,
    But here he paused and turned his head
    And looked across the Weald, and said
    It is a land worth fighting for.

2. Verena Smith in her introduction to the *Town Book of Lewes*, 1702-1837, states that among trade terms that of cordwainer was used before 1752 but surviving as would be expected in a provincial town some years after the more updated fashions were found in London. Most trades at that time had a very wide remit. Thus a flax dealer traded in sackcloth, cheesecloth, rope, thread, bed cords, reins, halters, plough traces, fishing line and shoe hemp. And G. F. Richardson sold silk and lace, furniture and was an undertaker.

3. Cowper's Poem 'The Cottager':

    Just knows and knows no more, her Bible true
    A truth the brilliant Frenchman never knew;
    And in that charter reads with sparkling eyes,
    Her tittle to a treasure in the skies.

4. Constables and headboroughs were elected annually, with legal authority and tax-collecting powers. They also established a night-watch and appointed: Ale Conner, Pound-keeper, Clerk of the Market, Town Cryer, Town Scavenger and Clock Keeper. Members of the Society of Twelve were forerunners of the jury system.

5. Algernon Sidney (1622-83) wounded at Marston Moor in 1644, entered parliament the next year and as an extreme republican rebelled against Cromwell's usurpation of power and retired to Penshurst, Kent. In 1683, he was implicated in the Rye House Plot to murder Charles II and James, Duke of York. The plot was foiled by the early departure of the royal pair from Newmarket and he was beheaded. John Wyndham, Lord Egremont, in his autobiography says that 'Algernon' comes from the Norman French meaning a whiskered one!

6. There is an unexplained entry in Gideon Mantell's journal dated 13 July 1820. 'Afterwards rode on to Mr Cornwell's near Cross in Hand; he was the only brother of my dear old lady Mrs C of Lewes, who fostered me when a child, and treated me with the greatest kindness while she lived.' Deborah Cadbury in *The Dinosaur Hunters* suggests that Mrs C. helped him financially when he went to London.

7. Lord Sheffield with his family and Thomas Pelham had first-hand knowledge of the French Revolution having visited Paris in 1791, attended meetings of the National Assembly and the Societé des Jacobins and listened to 'warm debates' about the king's future.
8. Non-Anglicans could not attend Oxford but they were permitted to become undergraduates at Cambridge, though denied the potential to sit for degrees.

9. A. H. Wilds succeeded his father as an architect but also in 1843 invented a patent for cleaning horizontal chimney flues in such a way as to obviate the necessity for using climbing boys.

10. Francesco Bartolozzi (1727-1815) was an Italian and a founder member of the Royal Academy. He had been invited to England by George III and appointed court engraver, remaining in England until 1802. He popularised the stipple process of colour engraving which by the absence of firm contours achieved an easy effect of elegance and sweetness. He made a small number of prints from historical paintings. With the foundation of the Royal Academy in 1768, Sir Joshua Reynolds made the purchase of pictures fashionable among the rising middle classes seeking a hallmark of gentility.

11. It is an irony that the eponym, Parkinsonism, is rarely attached to the agitated tremor of paralysis agitans, which has to be differentiated from Essential Tremor, and is more firmly attached to bradykinetic Parkinsonism, Post-encephalitic Parkinsonism and Progressive Supra-nuclear Paralysis, in all of which the predominant symptom is limitation and retardation of movement. As Charcot observed, the tremor and weakness of paralysis agitans may be slight or occur late in the course of the disease but for the majority of cases both symptoms are conspicuous.

12. It is a pity that this account came in 1830 some years after his discovery of the Iguanodon tooth.

FOSSILS

1. Bakewell's father was a well known stock breeder of sheep and longhorn Leicestershire Cattle. Bakewell himself was interested in the processing of wool before starting to give geological lectures. His association with Silliman began when Silliman read his *Introduction to Geology*.

2. Baron Cuvier, like the Mastodon, Megatherium and Mammoth, was himself megalocephalic. A post-mortem was performed on 15 May 1832 to contemplate the instrument of this powerful intelligence. The brain of Georges Cuvier weighed 1,830 grams, more than 400 grams above average and 200 grams larger than any non-diseased brain previously weighed. As a nobleman he had hidden himself in Normandy during the reign of terror before securing a lowly post at the Museum National d'Histoire comparing the anatomy of living creatures with those of the fossils and identifying an ancient sea lizard, Mosasaurus (lizard of the Meuse). In his sixties and hugely fat he earned the sobriquet, the Mammoth.

3. Mantell's memorial in St Michael's church, Lewes has a background of Sussex marble supporting a large brass plate with ornamental borders and a coat of arms in red enamel. There is a long inscription of over 200 words. It was originally placed below that to his father by his son Reginald in 1857.

4. Dr William Henry Fitton (1780-1861) described elsewhere by Mantell as a mad Irishman, was born in Dublin and studied the rocks in Ireland before those in Scotland and England. He qualified in Medicine in Edinburgh, also getting an MD from Cambridge and practised medicine until he married a wealthy wife. In 1836 he wrote on the strata between the chalk and the Oxford Oolite and separated lower and upper greensand from the gault. He became the energetic secretary of the Geological Society and clashed with the President, G. B. Greenough, over Smith's contribution to strata and tried unsuccessfully to get Smith elected to the Geological Society.

5. Sir Everard Home was Surgeon to the King and reputed to be Britain's leading anatomist. He had examined Mantell for the MRCS. He is remembered for plagiarising his brother-in-law, John Hunter's unpublished manuscripts, burning the originals. He raced into print to describe the Ichthyosaurus as a crocodile, delaying more scientific evaluation

LEWES

1. 'I purchased the three celebrated engravings of Bartolozzi of the resurrection of a pious family; they were framed and glazed. I gave £3 for the set. A supposed portrait of Oliver Cromwell, formerly belonging to the Duke of Newcastle was sold at the sale of Holland Place – £1. The age does not correspond with that of Oliver's and yet the countenance very much resembles the acknowledged portrait of the Protector.'

2. After the hangings the corpses were decapitated in public view by a masked figure whose speed and dexterity in severing the heads roused suspicions that he was a member of the medical profession with a newspaper article stating that he was a young surgeon from Argyll Street; all of which led erroneously to an arson attack, the stabbing and attempted murder of Thomas Wakley, founder of the medical journal, The Lancet.

3. An obituary of Gideon Mantell in the *Sussex Weekly Advertiser* of 1852 claimed that Gideon Mantell became a Licentiate of the Society of Apothecaries, 'a novelty in a country practice'. The Worshipful Society of Apothecaries have no record of Gideon Mantell becoming a Licentiate but confirms that of Joshua. (Dee Cook, Archivist, personal communication).

MUSEUM

1. Thomas Hodgkin (1798-1866) would be recognised today as the leading physician of that era. His name is used eponymously as Hodgkin's disease for lymphadenoma. He was a Quaker and social reformer, qualifying an an MD at Edinburgh and as a Licentiate of the Royal College of Physicians, London (LRCP). He refused the Fellowship of the College because non-Anglicans were not customarily allowed and he did not want exemption from a discretionary rule.

BRIGHTON

1. The obituary in *The Lancet* gives a very poor opinion of Sir Henry Halford (1761-1844). He was all tact and nothing else and opposed to the physical examination of patients, knew little of pathology and disliked innovation. Elsewhere he is described as vain, cringing and lengthy, His unprofessional gossip about his patients was shared in later years by Lord Moran in his book on Churchill.

2. Given that he had driven a hard bargain, had forsaken his un-gentrified patients, and spent days and weeks away from his practice in Lewes, their reaction was fully understandable.

3. The gifts he sent Yale College and to Silliman were Septaria from the pelitic sediments of the lias near Weymouth. Defined as limestone concretions, they tended to be spheroidal or elliptical, scored and grooved by radiating and polygonal patterns of calcite veins. They could be hung on a wall; though according to Mantell, tables of this kind are 'among the choicest pieces of furniture in the mansions of our nobility'. Needless to say, Mantell had one in his drawing room.

4. It is not surprising that Mantell should have been interested in the youthful poetry of his contemporary, William Cullen Bryant (1794-1878). As a vigorous opponent of slavery and an advocate of the new Republican Party, his poetry was distinctly political and philosophical. 'The Embargo' (1808) satirised Thomas Jefferson's government. He edited the *North American Review* in which his other poems including 'Thanatopsis' were published.

5. Charterism was a working-class radical movement seeking democratic rights for all men. The six points of the People's Charter (1838) were universal male suffrage, the abolition of property qualifications for MPs, parliamentary constituencies of equal size, a secret ballot, payment for MPs, and annual general elections. Petitions were presented to parliament in 1839 and 1841 but rejected by huge majorities.

6. Is it coincidental that visitor numbers to the British Museum rose substantially following the arrival of the Mantellian Collection?

1827-8	81,000
1837-8	266,000
1850	720,643

LONDON

1. In the third edition he quotes Southey:

> Go little book for this my solitude,
> I cast thee on the waters – go thy ways
> And if, as I believe, thy vein be good
> The world will find thee after many days.

SOIRÉE

1. Sir Woodbine Parish (1796-1881) was a diplomat with an interest in geology who served twice in Naples as well as in Buenos Aires. As a diplomat he was best known for the final text of the treaty of peace after the overthrow of Napoleon.

2. Alexander Gordon Melville was a comparative anatomist, co-author with Strickland of *The Dodo*, and later Professor of Zoology at Queen's University, Galway.

3. Fox and grapes – said of someone who wants a thing badly but cannot obtain it, and so tries to pretend that he or she does not really want it at all. The fox tried in vain to get the grapes, but when he found that they were beyond his reach, went away saying the grapes are sour.

4. William Parsons, Lord Oxnewtown, and later Earl of Rosse (1800-67) was a mathematician, astronomer and Member of Parliament who built The Leviathan of Parsontown, for six decades the largest telescope in the world, with a Newtonian reflector, a six-foot aperture and four-ton mirror.

FIFTIES

1. Reginald (1827-57) returned to England on his father's death to oversee the estate and then went to Allahabad in India where he was caught up in the Sepoy Mutiny and died of cholera. He, rather than Walter, preserved the last few pages of Mantell's journal.

HIS PERSONALITY AND HEALTH

1. In May 1842 he was confined to bed with 'neuralgia of the heart' and took 75 drops of laudanum and 12 drops of prussic acid before any relief was obtained. I suspect that this was an episode of palpitations rather than angina.

FAMILY TREES

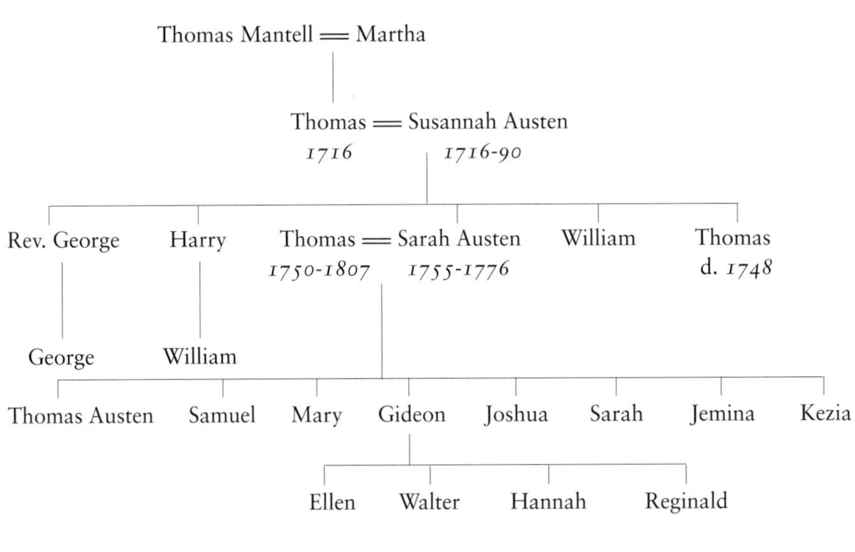

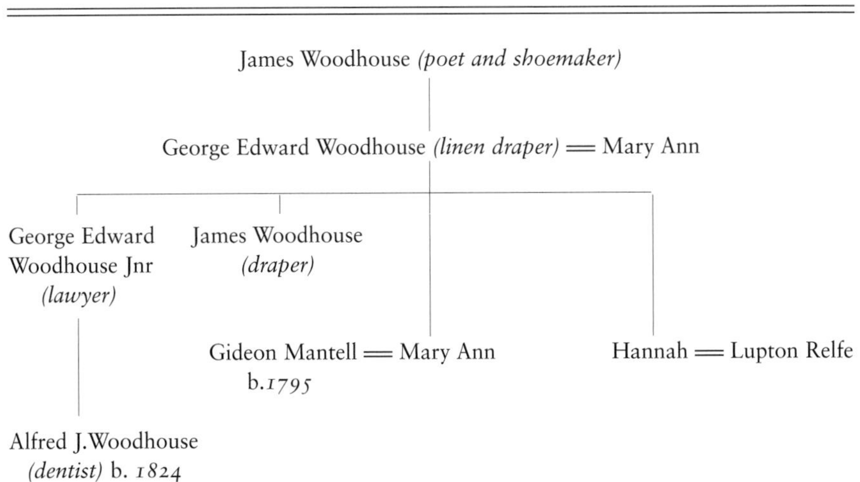

BIBLIOGRAPHY

PRINCIPAL WORKS

Brook, A., *Mantell Memoirs Series* (2 vols.), Lewes: Barbican Library, 2002.

Cadbury, D., *The Dinosaur Hunters*, London: Fourth Estate, 1988

Clark, P., *British Clubs and Societies. 1560-1800*, Oxford: Oxford University Press, 2000.

Critchley, E., *Hallucinations and their Impact on Art*, Preston: Carnegie Press, 1967.

Critchley, E., *Neurological Emergencies*, London: Saunders, 1988.

Critchley, E. and A. Eisen (eds.) *Spinal Cord Diseases*, London: Springer Verlag, 1997.

Curwen, E. C. (ed.) *The Journal of Gideon Mantell, 1818-52*, Oxford: Oxford University Press, 1940.

Dean, D., *Gideon Mantell and the Discovery of Dinosaurs*, Cambridge: Cambridge University Press, 1999.

Dean, D., *Gideon Algernon Mantell: A Bibliography with Supplementary Essays*, New York: Delmer, 1998.

Dellheim, C., *The Face of the Past: The Preservation of the Medieval Inheritance in Victorian England*, Cambridge: Cambridge University Press, 1982.

Egremont, Lord, *Wyndham and Children First*, Macintosh, 1968

Evans, J., *A History of the Society of Antiquities*, Oxford: Oxford University Press, 1986.

Fairbank, J. C. T., 'William Adams and the Spine of Gideon Algernon Mantell', *Spine*, 18: 909-913, 1993.

Fairbank, J. C. T., 'The Spinal injury of Gideon Mantell', *Annals of the Royal College of Surgeons of England*, 86: 349-352, 2004.

Freeman, M. *Victorians on the Prehistories: Tracks of a Former World*, London: Yale University Press, 2004.

Hays, J. N., 'The London Lecturing Empire 1800-50', in I. Inkster and J. Morrell (eds.) *Metropolitan and Provincial Science in British Culture*, London: Hutchinson, 1983.

Levine, P., *The Amateur and the Professional: Antiquities, Histories and Archaeologists in Victorian England, 1838-80*. Cambridge: Cambridge University Press, 1986.

Mantell, G. A., *Journal*, 4 vols. 1818-52. (Unpublished. Held at the Sussex Archaeological Society Library, Barbican House, Lewes, Sussex).

Mantell, G. A., *Memoirs of the Life of a Country Surgeon*, London, 1845.

Spokes, S., *Gideon Algernon Mantell LLD FRCS FRS*, London: Bale and Sons, 1927.

Spokes, S., 'A Case of Circumstantial Evidence', *Sussex County Magazine*: 118-122, 1937.

ADDITIONAL WORKS

Allen, R. C., *The British Industrial Revolution: A Global Perspective*, Cambridge: Cambridge University Press, 2009.

Bishop, W. J., 'The Evolution of the General Practitioner in England', in E. A. Underwood (ed.) *Science, Medicine and History: Essays on the Evolution of Scientific Thought and Medical Practice*, Oxford: Oxford University Press, 1953.

Brook, A., *Gideon Mantell – Memento Mori*, West Sussex Geological Society, 2002.

Brook, A., *The Fallacy of Mantell's Suicide*, West Sussex Geological Society, 2006.

Brook, A., The End of Mantell's Spine, West Sussex Geological Society, 2007.

Brent, C., *Pre-Georgian Lewes*, Lewes: Colin Brent Books, 2004.

Brent, C., *Georgian Lewes*, Lewes: Colin Brent Books, 2002.

Bynum, W. F., *Science and the Practice of Medicine in the Nineteenth Century*, Cambridge: Cambridge University Press, 1994.

Darwin, E., *Zoonomia*, Middlesex: Echo, 2007.

Dobson, J., 'The Anatomizing of Criminals', *Annals of the Royal College of Surgeons of England*, 9: 112-8, 1951.

Fairbank, J. C. T., 'Historical Perspective: William Adams, the Forward Bending Test, and the Spine of Gideon Algernon Mantell', *Spine*, 29: 1953-55, 2004.

Fairbank, J. C. T., 'The Spine of Gideon Mantell', *Annals of the Royal College of Surgeons of England*, 86: 349-57, 2004.

Fleming, B., *Lewes: Two Thousand Years of History*, Seaford: SB Publications, 1994.

Gilbert, E. W., *Brighton: Old Ocean's Bauble*, London: Methuen, 1954.

Gribbin, J., *History of Western Science 1543-2001*, London: Folio Society, 2006.

Hilton, B., *A Mad, Bad, and Dangerous People? England 1783-1846*, Oxford: Clarendon Press, 2006.

Hodgkin, T. & W. M. Adams, 'A Case of Distortion of the Spine with Observations on Rotation of the Vertebrae as a Complication of Lateral Curvature', *Medico-Chirurgical Transactions*, 37: 167-180, 1854.

Holmes, G., *Augustan England: Professions, State and Society, 1680-1730*, London: Allen & Unwin, 1982.

Horsfield, T. W., *The History and Antiquities of Lewes and its Vicinity*, Lewes: J. Baxter, 1824.

Huxley, R. (ed.), *The Great Naturalists*, London: Thames & Hudson, 2007.

Irvine, W., *Apes, Angels and Victorians*, London: McGraw Hill, 1955.

Jones, R., 'Thomas Wakley, Plagiarism, Libel and the Founding of *The Lancet*', *Journal of the Royal Society of Medicine*: 404-10, 2009.

Keane, J., *Tom Paine: A Political Life*. London: Bloomsbury, 1995.

Loudon, I., *Medical Care and the General Practitioner, 1750-1852*, Oxford: Clarendon Press, 1986.

Lower, M. A., *The Worthies of Sussex*, Lewes, 1865.

Morris, A. D., 'G. A. Mantell', *Proceedings of the Royal Society of Medicine*, 65: 215-21, 1972.

Ryan, F., *Darwin's Blind Spot: Evolution Beyond Natural Selection*, London: Houghton Mifflin, 2003.

Pettigrew, T. J., 'John Abernethy', in R. Coope (ed.), *The Quiet Art: A Doctor's Anthology*, London: Livingstone, 1988.

Porter, R., *English Society in the Eighteenth Century*, London: Penguin, 1990.

Rupke, N. A., *The Great Chain of History*, Oxford: Clarendon Press, 1983.

Smith, V., *Town Book of Lewes, 1702-1837*, Lewes: Sussex Record Society, 1978.

Torrens, H., 'When did the dinosaur get its name?', *New Scientist*, 1815: 40-44, 1992.

Torrens, H., 'Politics and Palaeontology', in J. Farlow & M. K. Brett-Surman (eds.), *The Complete Dinosaur*, Indiana: Indiana University Press, 1999.

Trevelyan, G. M., *English Social History*, London: Longmans Green, 1946.

Trevelyan, G. M., *British History in the Nineteenth Century and After*, London: Longmans Green, 1947.

Unglow, J., *The Lunar Men: Five Friends Whose Curiosity Changed The World*, London: Faber & Faber, 2002.

Waddington, K., *Medical Students and Medical Education at St Bartholomew's* (Unpublished), 2001.

Weinkove, R., 'The Rise and Fall of the Medicinal Leech', *Student BMJ*, August, 1998.

Woodward, E. L., *The Age of Reform, 1815-1870*, Oxford: Clarendon Press, 1996.

INDEX